AF360739

Protein Adsorption and Conformational Changes

Protein Adsorption and Conformational Changes

Editor

Michael Assfalg

MDPI • Basel • Beijing • Wuhan • Barcelona • Belgrade • Manchester • Tokyo • Cluj • Tianjin

Editor
Michael Assfalg
Department of Biotechnology
University of Verona
Verona
Italy

Editorial Office
MDPI
St. Alban-Anlage 66
4052 Basel, Switzerland

This is a reprint of articles from the Special Issue published online in the open access journal *Molecules* (ISSN 1420-3049) (available at: www.mdpi.com/journal/molecules/special_issues/Protein_Adsorp_ Conform).

For citation purposes, cite each article independently as indicated on the article page online and as indicated below:

LastName, A.A.; LastName, B.B.; LastName, C.C. Article Title. *Journal Name* **Year**, *Volume Number*, Page Range.

ISBN 978-3-0365-2691-1 (Hbk)
ISBN 978-3-0365-2690-4 (PDF)

Contents

About the Editor

Michael Assfalg

Michael Assfalg (MA) earned his PhD in Chemistry from the University of Florence. He was then a post-doctoral research assistant at the University of Maryland and then at the Biozentrum, University of Frankfurt, where he performed research in the field of biomolecular NMR spectroscopy. MA is currently an Associate Professor of Organic Chemistry at the University of Verona (UNIVR). He is a teacher of Organic Chemistry and Supramolecular Chemistry. He is a member of the steering committee of the PhD Course in Biotechnology and Coordinator of the master's and bachelor's degrees in Biotechnology at UNIVR. MA received funding as P.I. under local and national schemes, which allowed him to establish his own research group. MA is interested in elucidating the structural and binding properties of chemical entities with biological function, including metabolites, biopolymers, supramolecular assemblies, and synthetic receptors.

MDPI

Editorial

Protein Adsorption and Conformational Changes

Michael Assfalg 

Department of Biotechnology, University of Verona, 37134 Verona, Italy; michael.assfalg@univr.it

Protein adsorption onto surfaces of diverse materials of both natural and artificial origin is of utmost relevance in many areas of research and technology: medicine, pharmaceutical sciences, analytical sciences, biotechnology, nanotechnology, and cell biology, among others [1–3]. In general, any process involving an interface in which contact with a protein solution occurs is likely to be influenced by protein adsorption to the interface [4]. However, despite considerable advances in the field [5,6], our understanding of the mechanisms and consequences of adsorption remains incomplete for a number of reasons. Firstly, the vast diversity of material surfaces makes it impossible to describe, in detail, all the biophysicochemical interactions at play, preventing us from disclosing general paradigms. Secondly, the highly heterogeneous and often dynamic character of interfaces poses tremendous challenges to both experimental and computational approaches, which therefore often provide only partial descriptions. Thirdly, proteins are not uniform, static entities with homogeneous surface properties; instead, they exhibit protein-specific plasticity, complex shapes, and anisotropic surface features.

Proteins adsorb in differing numbers, densities, orientations, and conformations, depending on the chemical and physical properties of the surface and their inherent molecular characteristics [7]. The nature of protein interactions with surfaces ranges from weak Van-der-Waals interactions to strong electrostatic attraction, resulting in a wide range of binding affinities and exchange kinetics. It is well accepted that the properties of surfaces will govern both the degree of conformational change and the orientation of nonspecifically adsorbed proteins [8]. Because of the participation of a significant portion of the protein, adsorption may induce conformational changes from the solution state [9]. Perturbed conformations may result in the impairment of function and exposure of cryptic epitopes that can produce unintended effects in biological signaling [10]. Furthermore, adsorption-induced rearrangements and crowding effects may facilitate protein-protein interactions and aggregation. The latter issue has been intensely investigated, particularly in relation to protein misfolding and amyloid formation. Therefore, molecular-level descriptions of protein adsorption would significantly contribute to our understanding of the phenomenon [11], and are essential for the effective control of protein–surface interactions.

To rationalize the way in which proteins interact with surfaces, it is useful to consider the peculiarities of the different types of surfaces. For example, it is known that nanomaterials exhibit unique properties that differ from those of the bulk substances, and are therefore attracting considerable interest in many areas of medicine and technology. As opposed to planar surfaces, nanoparticles have a very large surface-to-volume ratio, so that even small amounts of particles present extremely large surface areas available for protein binding. Moreover, nanoparticle surface curvature has a profound influence on the adsorption process. Thus, protein-nanoparticle interactions have been extensively investigated, and it is not surprising that they are discussed in a number of reports in this Special Issue [12–15]. Lipid surfaces are instead the subject of another contribution in this issue [16]. Lipid layers are indeed important mimics of biological membranes and may be employed to assemble solid-supported membranes, which are well-suited for studying conformational dynamics phenomena of peripheral or integral membrane proteins.

In this Issue, Perera et al. [12] describe a systematic study of peptide and protein adsorption to PEGylated gold nanoparticles. PEGylation is a universally adopted strategy for the passivation of nanoparticles, which reduces unintended protein adsorption and extends the lifetime of gold nanoparticles in biofluids. Systematic investigations are crucial to establish quantitative relationships between molecular/particle properties and binding propensities. The authors use an NMR spectroscopy-based approach, which allows them to obtain the in situ quantification of protein binding and the determination of association constants and kinetic parameters. The study reveals important determinants of the differing penetration of test proteins into PEG coatings of variable density and thickness.

Protein interaction with gold nanoparticles is also the subject discussed in another original report of this issue, contributed by Hunashal et al. [13], although in this case, the focus is on the influence of nanoparticles on protein fibrillogenesis. The authors introduce a method of investigation of site-specific transient interactions based on NMR signal perturbations induced by an extrinsic paramagnetic agent. Their study involves citrate-stabilized gold nanoparticles and β2-microglobulin, a paradigmatic example of amyloidogenic protein. Paramagnetic perturbation mapping provides insight into protein regions which become incompetent for protein–protein interactions in the presence of nanoparticles, thereby contributing to explain why these nanoobjects inhibit the fibrillogenesis of this protein.

Alpha-synuclein is another prototypical amyloidogenic protein, associated with a number of neurodegenerative conditions, including Parkinson's disease. Different from β2-microglobulin, alpha-synuclein is intrinsically disordered in solution, but undergoes conformational rearrangement in pathological conditions, or upon proper stimulation in vitro. The review by D'Onofrio et al. [15] summarizes a large number of studies that have been conducted to understand how the unstructured polypeptide adapts to nanoparticle surfaces, and how its adsorption influences protein self-assembly and formation of fibrillar structures. Indeed, nanoparticles represent an attractive alternative to traditional drugs for the development of agents capable of mitigating aberrant protein aggregation in devastating brain diseases. The conformational plasticity of alpha-synuclein further offers interesting opportunities for the development of new nanobiomaterials displaying emergent properties.

Georgieva reviewed the current progress of research based on the use of electron paramagnetic resonance (EPR) to study the protein conformational dynamics of proteins associated with surfaces of membranes or nanomaterials [14]. The author describes how EPR in its continuous-wave and pulse spectroscopy modalities, in combination with spin labeling, provides information on local dynamics and long-range conformational rearrangements of proteins. The reported case studies include investigations of the disordered, amyloidogenic proteins tau and alpha-synuclein on interaction with lipid membranes, which contributed significant insights into the structural changes experienced by the proteins a upon transition from the solution to the lipid layer. The review further considers examples of application of the technique to investigate the conformational behavior of proteins immobilized through adsorption onto surfaces of engineered nanomaterial particles and synthetic membranes, at the basis of new technological developments in biotechnology and analytical sciences.

Protein immobilization on engineered membranes is indeed a convenient preparative procedure for investigations of membrane protein function. Tadini-Buoninsegni reviewed research articles focused on applications of an electrophysiological method based on solid supported membranes (SSMs) to elucidate the ion transport mechanism in P-type ATPases, a superfamily of membrane transporters. During the catalytic cycle, the ATPases typically undergo structural rearrangements and conformational transitions to perform ion transport across the membrane. Membrane layers or vesicles that embed the transport protein are adsorbed on the SSM. The proteins are activated by a substrate concentration jump at the SSM, and the charge translocations during the reaction cycle are monitored by measuring a transient current signal. The described approach has a broad scope, and also finds application in drug discovery, due to its ability to monitor protein/drug interactions.

One could envision that the study of protein/nanoparticle interactions on SSMs could be another field of interest in the future.

In summary, protein adsorption to surfaces plays a crucial role in both basic and applied research. The reports contained in this Special Issue show that significant progress has been made in our understanding of the adsorption mechanisms, and of complex conformational dynamics at interfaces, contributing to set the basis for future developments.

I wish to thank the authors for their contributions to this Special Issue, all the reviewers for their invaluable work, and the editorial staff of *Molecules*, particularly Assistant Editor Emity Wang for her kind assistance during the preparation of this Special Issue.

Conflicts of Interest: The authors declare no conflict of interest.

References

1. Schmidt, D.R.; Waldeck, H.; Kao, W.J. Protein Adsorption to Biomaterials. In *Biological Interactions on Materials Surfaces*; Puleo, D.A., Bizios, R., Eds.; Springer: New York, NY, USA, 2009; pp. 1–18, ISBN 978-0-387-98160-4.
2. Frutiger, A.; Tanno, A.; Hwu, S.; Tiefenauer, R.F.; Vörös, J.; Nakatsuka, N. Nonspecific Binding—Fundamental Concepts and Consequences for Biosensing Applications. *Chem. Rev.* **2021**, *121*, 8095–8160. [CrossRef] [PubMed]
3. Lynch, I.; Dawson, K.A. Protein-Nanoparticle Interactions. *Nano Today* **2008**, *3*, 40–47. [CrossRef]
4. Brash, J.L.; Horbett, T.A. (Eds.) *Proteins at Interfaces: Physicochemical and Biochemical Studies*; ACS Symposium Series; American Chemical Society: Washington, DC, USA, 1987; Volume 343, ISBN 978-0-8412-1403-3.
5. Nel, A.E.; Mädler, L.; Velegol, D.; Xia, T.; Hoek, E.M.V.; Somasundaran, P.; Klaessig, F.; Castranova, V.; Thompson, M. Understanding Biophysicochemical Interactions at the Nano–Bio Interface. *Nat. Mater.* **2009**, *8*, 543–557. [CrossRef] [PubMed]
6. Rabe, M.; Verdes, D.; Seeger, S. Understanding Protein Adsorption Phenomena at Solid Surfaces. *Adv. Colloid Interface Sci.* **2011**, *162*, 87–106. [CrossRef] [PubMed]
7. Sigal, G.B.; Mrksich, M.; Whitesides, G.M. Effect of Surface Wettability on the Adsorption of Proteins and Detergents. *J. Am. Chem. Soc.* **1998**, *120*, 3464–3473. [CrossRef]
8. Roach, P.; Farrar, D.; Perry, C.C. Interpretation of Protein Adsorption: Surface-Induced Conformational Changes. *J. Am. Chem. Soc.* **2005**, *127*, 8168–8173. [CrossRef] [PubMed]
9. Ta, T.C.; McDermott, M.T. Mapping Interfacial Chemistry Induced Variations in Protein Adsorption with Scanning Force Microscopy. *Anal. Chem.* **2000**, *72*, 2627–2634. [CrossRef] [PubMed]
10. Lynch, I.; Dawson, K.A.; Linse, S. Detecting Cryptic Epitopes Created by Nanoparticles. *Sci. Signal.* **2006**, *327*, pe14. [CrossRef] [PubMed]
11. Penna, M.J.; Mijajlovic, M.; Biggs, M.J. Molecular-Level Understanding of Protein Adsorption at the Interface between Water and a Strongly Interacting Uncharged Solid Surface. *J. Am. Chem. Soc.* **2014**, *136*, 5323–5331. [CrossRef] [PubMed]
12. Perera, Y.R.; Xu, J.X.; Amarasekara, D.L.; Hughes, A.C.; Abbood, I.; Fitzkee, N.C. Understanding the Adsorption of Peptides and Proteins onto PEGylated Gold Nanoparticles. *Molecules* **2021**, *26*, 5788. [CrossRef] [PubMed]
13. Hunashal, Y.; Cantarutti, C.; Giorgetti, S.; Marchese, L.; Fogolari, F.; Esposito, G. Insights into a Protein-Nanoparticle System by Paramagnetic Perturbation NMR Spectroscopy. *Molecules* **2020**, *25*, 5187. [CrossRef] [PubMed]
14. Georgieva, E.R. Protein Conformational Dynamics upon Association with the Surfaces of Lipid Membranes and Engineered Nanoparticles: Insights from Electron Paramagnetic Resonance Spectroscopy. *Molecules* **2020**, *25*, 5393. [CrossRef] [PubMed]
15. D'Onofrio, M.; Munari, F.; Assfalg, M. Alpha-Synuclein—Nanoparticle Interactions: Understanding, Controlling and Exploiting Conformational Plasticity. *Molecules* **2020**, *25*, 5625. [CrossRef] [PubMed]
16. Tadini-Buoninsegni, F. Protein Adsorption on Solid Supported Membranes: Monitoring the Transport Activity of P-Type ATPases. *Molecules* **2020**, *25*, 4167. [CrossRef] [PubMed]

molecules

Article

Understanding the Adsorption of Peptides and Proteins onto PEGylated Gold Nanoparticles

Yasiru Randika Perera [1,†], Joanna Xiuzhu Xu [1,†], Dhanush L. Amarasekara [1], Alex C. Hughes [1], Ibraheem Abbood [2] and Nicholas C. Fitzkee [1,*]

[1] Department of Chemistry, Mississippi State University, Starkville, MS 39762, USA; yrp7@msstate.edu (Y.R.P.); xx79@msstate.edu (J.X.X.); dla216@msstate.edu (D.L.A.); ach436@msstate.edu (A.C.H.)

[2] Department of Chemistry, University of Arkansas at Little Rock, Little Rock, AR 72204, USA; ibraheemabbood12@gmail.com

* Correspondence: nfitzkee@chemistry.msstate.edu; Tel.: +1-662-325-1288

† These authors contributed equally to this work.

Abstract: Polyethylene glycol (PEG) surface conjugations are widely employed to render passivating properties to nanoparticles in biological applications. The benefits of surface passivation by PEG are reduced protein adsorption, diminished non-specific interactions, and improvement in pharmacokinetics. However, the limitations of PEG passivation remain an active area of research, and recent examples from the literature demonstrate how PEG passivation can fail. Here, we study the adsorption amount of biomolecules to PEGylated gold nanoparticles (AuNPs), focusing on how different protein properties influence binding. The AuNPs are PEGylated with three different sizes of conjugated PEG chains, and we examine interactions with proteins of different sizes, charges, and surface cysteine content. The experiments are carried out in vitro at physiologically relevant timescales to obtain the adsorption amounts and rates of each biomolecule on AuNP-PEGs of varying compositions. Our findings are relevant in understanding how protein size and the surface cysteine content affect binding, and our work reveals that cysteine residues can dramatically increase adsorption rates on PEGylated AuNPs. Moreover, shorter chain PEG molecules passivate the AuNP surface more effectively against all protein types.

Keywords: NMR spectroscopy; gold nanoparticles; PEGylation; adsorption; passivation

check for
updates

Citation: Perera, Y.R.; Xu, J.X.; Amarasekara, D.L.; Hughes, A.C.; Abbood, I.; Fitzkee, N.C. Understanding the Adsorption of Peptides and Proteins onto PEGylated Gold Nanoparticles. *Molecules* **2021**, *26*, 5788. https://doi.org/10.3390/molecules26195788

Academic Editor: Michael Assfalg

Received: 29 July 2021
Accepted: 20 September 2021
Published: 24 September 2021

Publisher's Note: MDPI stays neutral with regard to jurisdictional claims in published maps and institutional affiliations.

1. Introduction

When nanoparticles (NP) encounter a biological environment, biomolecules will spontaneously adsorb to the NP surface, forming a biomolecular corona. For many biomedical applications, the NP surfaces are designed to elicit the adsorption of biomolecules for bioimaging and biosensing [1]. On the other hand, some NP surfaces are fabricated to limit the adsorption of biomolecules, preventing recognition by the human body [2]. This is especially true for gold nanoparticles (AuNPs) that must pass through the blood–brain barrier or otherwise avoid clearance from the immune system [3,4]. However, when administered intravenously [5], AuNPs are rapidly removed from the circulation and accumulate mainly in the liver and the spleen due to the opsonization and recognition by phagocytes [3,6]. Many studies attempt to reduce the undesirable uptake of AuNPs by modifying the physicochemical properties of AuNP such as size, surface charge, hydrophilicity, and surface functionality [4,7–11].

A well-known method to avoid phagocytosis is to functionalize NP surfaces with polyethylene glycol (PEG), also called PEGylation [12,13]. There are numerous studies indicating that PEG-coated AuNPs reduce the protein corona formation and increase the retention time of the NPs in the circulatory system [14–16]. The passivating effect of PEG on AuNPs is thought to be conferred due to a reduced tendency for protein binding to a PEGylated surface [17,18]. As this biocorona layer of bound proteins is reduced, the immune system does not recognize the AuNPs, and therefore PEGylated NPs will persist [19,20]. In addition, PEG-passivated NPs are widely used as drug carriers as the solution exposed termini can be modified with a drug [21–24]. As an example, doxorubicin can be effectively attached to PEG-coated AuNPs to target cancer cells for treatment [25].

In practice, there are numerous approaches to PEGylation, and size, terminal modification, and charge can all be modified [26–28]. PEGylation is conducted using several unique techniques including detritylating PEG derivatives, covalent attachment, entrapment, or AuNP surface adsorption of PEG chains. The effect of the PEGylation depends on the PEG molecular weight, polymer chain architecture, and PEG surface density on the NP surface [29,30]. For example, a high density of PEG favors an extended "brush" conformation, which is more effective at passivation than the "mushroom" conformation adopted during less dense PEGylation [16]. This improved passivation likely results from a shift in PEG dynamics, with more constrained motions favoring improved passivation [31]. These findings are of great relevance for studying protein adsorption onto PEGylated spherical nanoparticle surfaces, and the protein adsorption resistance on the NP surfaces can vary drastically depending on the PEGylation strategy.

In addition to the properties of PEG itself, protein properties are also likely very important in determining the effectiveness of passivation, and some proteins are better than others at circumventing passivation strategies. For example, while PEGylation reduces nonspecific binding overall, gel electrophoresis and mass spectrometry identify differences in how individual proteins interact with poly(DL-lactide-co-glycolide) (PLGA) nanoparticles upon passivation with PEG [32]. Another recent study by Blume et al. used the protein corona of PEGylated superparamagnetic iron oxide NPs as a probe for proteomics analysis, demonstrating that PEG does not block binding from all proteins equally [33]. Both of these studies were carried out at shorter incubation periods, under 12 h, in which PEG exhibits excellent passivating characteristics; nevertheless, some proteins are clearly able to bind PEGylated NPs. At longer timescales, the problem of protein binding becomes more pronounced, especially at low PEG densities [34,35]. Some nanoparticle formulations persist for > 24 h in the body [36], and therefore characterizing how blood proteins interact with PEG-coated NPs is an important aspect of drug delivery optimization. In particular, understanding how different biomolecular properties influence PEGylated NP binding could lead to better passivation strategies, or it could reveal insight into why some NP systems resist passivation.

In this study, we have quantitatively characterized several different peptides and proteins as they adsorb to PEGylated AuNPs. Both small peptides and larger proteins are studied, and the presence of cysteine residues is also systematically investigated. The AuNPs are coated with 5K, 10K, and 30K thiolated PEG, allowing us to probe different levels of PEG coverage. Using NMR spectroscopy [37–39], we have quantified the kinetics of and final stoichiometry of biomolecular adsorption to PEGylated AuNPs, and we discuss trends and physical properties that lead to efficient binding to PEGylated AuNP surfaces.

2. Results

2.1. PEGylation of Gold Nanoparticles and Protein Selection

Prior to protein adsorption, AuNPs were incubated with an excess amount of different thiol PEG compounds (PEG-SH) to fully passivate the surface. The attachment of PEG-SH on AuNP was characterized using UV-Vis spectroscopy, Zeta potential and dynamic light scattering (DLS) (Table 1). An increasing hydrodynamic diameter was observed when larger PEG-SH molecules are bound to the AuNP surface. Zeta potentials of all AuNP-PEG conjugates increase from -38 ± 3 mV to around -20 mV, indicating the negatively charged citrate-capped AuNP surfaces are all saturated with PEGs. Transmission electron microscopy (TEM) is consistent with 15-nm AuNPs, showing no AuNP aggregation during synthesis of AuNP-PEGs. Visualization of the PEG in TEM is difficult due to the formvar grids used in TEM experiments, and the PEG layer itself was not consistently observed (Figure S2, Supporting Information). The localized surface plasmon resonance

(LSPR) peak of AuNPs at 520 nm has a decreasing degree of redshift of 3, 2, and 1 nm for 5K-PEG-SH-, 10K-PEG-SH-, and 30K-PEG-SH-coated AuNP, respectively (Table 1 and Figure S1, Supporting Information). This decrease in LSPR redshift as PEG size increases was previously reported [40], and the larger redshift for shorter PEGs likely arises from a mushroom-like structure of PEG: the end with a terminal thiol group binds to AuNP through a strong gold-thiolate bond, and the long PEG polymer chain adopts a random coil shape [14,40]. Based on this model, it is hypothesized that shorter PEG-SHs are grafted more densely on the AuNP than longer ones, which explains why PEG-SH-5K can most effectively change the dielectric constant of the immediate medium surrounding AuNP surface, and therefore induces the largest LSPR redshift. Indeed, the experimentally measured PEG density on these AuNP-PEG conjugates decreases from 0.96 ± 0.01 to 0.57 ± 0.01 PEG/nm^2 as PEG size increases from 5K to 30K (Table 1), supporting the predicted PEG conformational model on the AuNP surface. This is consistent with prior studies examining the trend of PEG-SH coverage on AuNPs [34,41]. Scheme 1 depicts the PEG-SH conformations on AuNP surface and PEG density decreases with larger PEG-SHs.

Table 1. The hydrodynamic diameter (D_H), localized surface plasmonic resonance (LSPR), Zeta potential and PEG density of AuNP and its three different conjugates of PEG.

AuNP-PEG Conjugate	D_H (nm)	Zeta Potential (mV)	LSPR Peak (nm)	PEG Density (PEG/nm^2)
AuNP	18.4 ± 1.1	-38 ± 3	520	0
AuNP-PEG-5K	27.7 ± 0.5	-16.4 ± 0.8	523	0.96 ± 0.01
AuNP-PEG-10K	39 ± 2	-22.1 ± 0.6	522	0.67 ± 0.01
AuNP-PEG-30K	78 ± 3	$-18 + 2$	521	0.57 ± 0.01

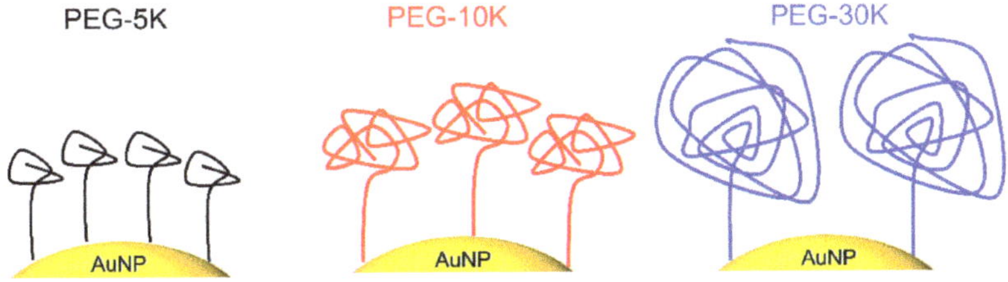

Scheme 1. Conformational model of PEG-SH binding to an AuNP surface.

Permeability of a biomolecule into the PEG chains is essential for its binding onto AuNP surface. In this work, protein size and surface thiol groups are hypothesized to be the key factors that determine permeability. This is for two reasons: First, smaller molecules are expected to diffuse into PEG chains more easily than larger ones. Second, thiol groups have a particularly strong affinity towards AuNP, which facilitates binding through formation of gold-thiolate bond [42,43]. Therefore, representative molecules are selected to be glutathione (GSH), the H1.5 peptide (with and without cysteine), wild-type GB3, the K19C GB3 variant, and bovine serum albumin (BSA) (Figure 1, Table 2). GSH is the smallest construct with only one thiol group, whereas the H1.5 peptide, which is phosphorylated and derived from a histone H1.5 segment, lacks a thiol. H1.5-Cys is identical to H1.5, but it contains a Cys residue at the 9th position. Larger protein candidates include WT GB3, which lacks cysteine residues, and K19C GB3, which contains one cysteine at position 19. BSA is the largest protein, and while it contains 35 Cys residues, only one of these is readily available to bond; the rest form disulfide bonds and are buried within the folded structure of the protein. The estimated net charge of each protein, assuming model compound pK$_a$ values, is also presented. PEGylated AuNPs retain a negative ζ-potential, suggesting that more basic proteins may be favored [44]. Although the proteins selected do not exhaustively probe protein charge, the H1.5 peptide is highly basic and could potentially indicate whether electrostatic interactions influence how efficiently proteins can permeate a PEG layer.

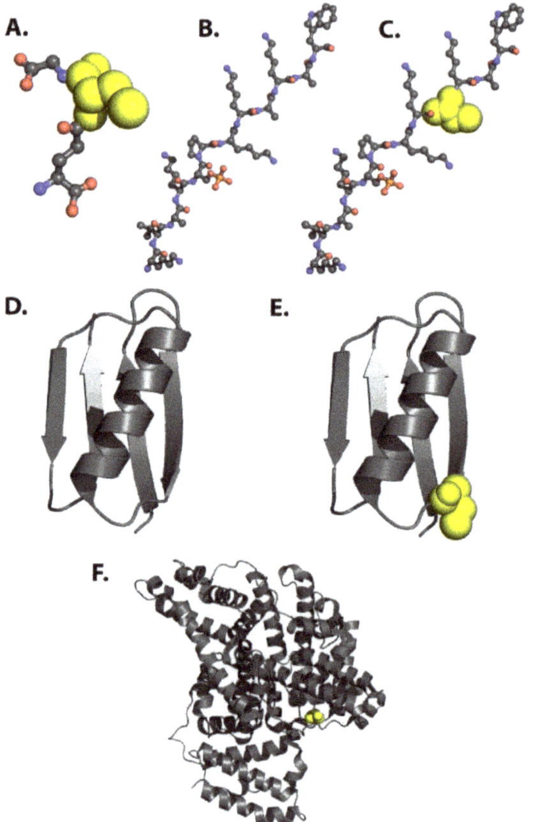

Figure 1. Peptides and proteins used in this study. (**A**) Glutathione peptide (GSH), (**B**) H1.5 peptide, (**C**) H1.5-Cys peptide, (**D**) Wild-type GB3 protein (WT GB3, PDB 2OED), (**E**) K19C GB3 variant (K19C GB3), and (**F**) bovine serum albumin (BSA, PDB 3V03 chain A). Peptides employ a ball-and-stick representation, whereas proteins are represented with cartoons. Cysteine residues that do not form intramolecular disulfide bonds are shown as yellow spheres in space filling representation. Note that peptides and proteins are not drawn to scale.

Table 2. Characterization of size, number of surface cysteines and net charge of the ligands used in this work.

	GSH	H1.5	H1.5-Cys	WT GB3	K19C GB3	BSA
Size (kDa)	0.3	1.4	1.4	6.2	6.2	66
Surface cysteines	1	0	1	0	1	1
Estimated net charge at pH 7	-1	5	5	-2	-3	-17

2.2. Adsorption of Small Peptides onto PEGylated AuNPs

We first monitored the adsorption of three small peptides onto PEG-SH coated AuNPs, and their adsorption rate constants and adsorbed amount are obtained for comparison (Figure 2A,B). The PEG size affects the adsorption rate constants as well as the final amount of peptides adsorbed onto AuNPs. All peptides are adsorbed fastest and in the greatest amount for AuNPs coated with 30K-PEG-SH, with adsorption rate and amount decreasing with smaller PEG size (Figure 2C,D). This indicates these molecules can penetrate more easily into the PEG layers formed with larger PEGs, and the AuNPs functionalized with larger PEGs have more available surface area for additional ligand binding. This observation is consistent with the grafting density of PEG-SH on AuNP, which is limited to an extent by the size of the coil-shaped PEG chains. The use of larger PEG-SH molecules results in less densely grafted PEG on AuNP, which facilitates diffusion of ligands into the PEG chains and leaves more unoccupied AuNP surface for binding [34]. Another observation is that GSH is adsorbed faster and in a larger amount than H1.5 on each type of AuNP conjugate. This is attributed to the smaller size of GSH (0.3 kDa) as compared to H1.5 (1.4 kDa), and high affinity afforded by a thiol group in GSH in contrast to no thiol group in H1.5.

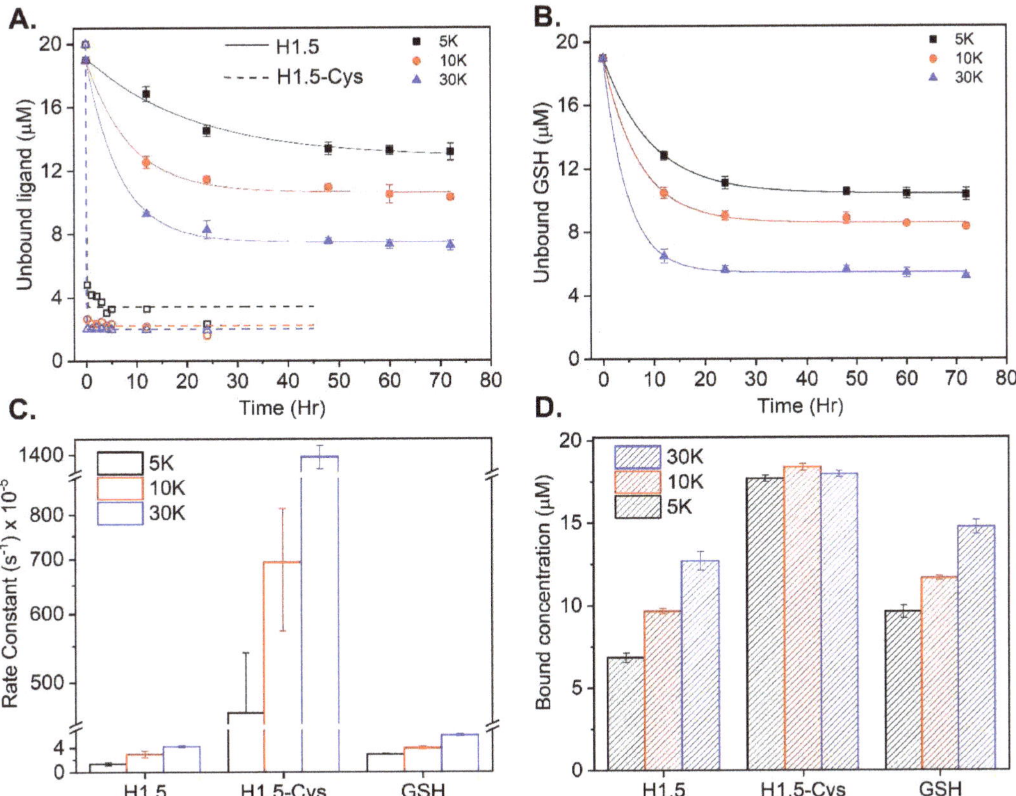

Figure 2. Adsorption of small peptides onto PEGylated AuNPs. (**A**) Free H1.5 (closed symbols, solid lines) and H1.5-Cys (open symbols, dashed lines) concentrations vs. time in the presence of 5K, 10K, and 30K PEGylated AuNPs during the 0–72 h incubation period. The kinetic study for H1.5-Cys was performed for only 24 hrs due to its fast adsorption kinetics. (**B**) GSH adsorption to PEGylated AuNPs during the same time duration. The curves in (**A,B**) are a pseudo-first-order kinetics fit, and the error bars are the standard error of the mean. (**C**) Adsorption rate constants for H1.5 and GSH for different PEGylated AuNPs obtained from the pseudo-first-order kinetic fit. (**D**) Bound concentration at the final time point of H1.5 and GSH for different PEGylated AuNPs after 72-hour incubation. The apparent adsorption rates and bound concentrations of all ligands used in this work are summarized in Table S1 (Supporting Information). Error bars in panels (**C,D**) represent uncertainties calculated from a weighted least squares minimization of the reduced chi-squared performed in OriginPro 2021b (Northampton, MA, USA).

Interestingly, incorporating a Cys residue into the H1.5 peptide dramatically enhances this peptide's affinity toward PEGylated AuNPs (open symbols, Figure 2A). H1.5-Cys is identical to H1.5, except that it contains a cysteine residue in the 9th position (A9C). In all of our experiments, near-complete binding of H1.5-Cys occurred within fifteen min—the dead time of our measurements (Figure 2A). While apparent rate constants could be fit for H1.5-Cys, these are lower-bound estimates, and binding of H1.5-Cys could be faster. In addition, the amount of H1.5-Cys bound was higher than H1.5 for all sizes of PEGylated AuNPs (Table S1, Figure 2D). Apparently, the H1.5-Cys peptide is able to readily penetrate the PEG layer and form a stable thiolate bond. Nearly all of the H1.5-Cys peptide can be adsorbed, suggesting that PEG is not a particularly good passivating compound for small, basic peptides containing thiol groups.

All of these experiments were performed under conditions with no added salt, so screening of electrostatic interactions should be minimal. The balance of size, electrostatics, and the presence of a thiol can be understood using the following simple scheme (Scheme 2):

$$Peptide + AuNP \underset{k_{-1}}{\overset{k_1}{\rightleftarrows}} Peptide\ in\ PEG\ Layer \overset{k_2}{\rightarrow} Bond\ Formed$$

Scheme 2. Kinetic scheme for peptide interaction with PEGylated AuNPs.

In this scheme, the peptide is able to penetrate the PEG layer in a fast-equilibrium process ($K_1 = k_1/k_{-1}$). This equilibrium is altered by electrostatics, which will favor penetration into the PEG layer for more basic peptides such as H1.5 (increasing K_1). It is feasible that the rate constant k_1 is sensitive to the peptide and to the density of the PEG layer. For example, smaller peptides (such as GSH) are able to penetrate the PEG layer more rapidly (increasing k_1), and denser PEG layers will likely decrease the speed of penetration (reducing k_1). In our work, k_2 is primarily dependent on whether a gold-thiolate bond can occur between the AuNP surface and a Cys residue. Other types of bonds are possible between proteins and Au surfaces, including amines and carboxylates [45–48], but the presence of a Cys residue should strongly increase the rate k_2. Here, our measured rate constants (k_{obs}) are approximately equal to $k_{obs} \approx K_1 k_2$, a scenario that is analogous to EX2 conditions for hydrogen exchange ($k_{-1} \gg k_2$) [49]. Elucidating specific values for K_1 and k_2 is challenging, but our data qualitatively support this scheme. H1.5 is highly basic (larger K_1), but it is larger than GSH (slower k_1) and lacks a thiol (slower k_2), therefore it exhibits the slowest adsorption kinetics. H1.5-Cys contains a thiol (fast k_2), and therefore it binds very quickly, within the first 15 min of mixing. GSH, which is acidic (weaker K_1) and contains a thiol (fast k_2), is intermediate between H1.5-Cys and H1.5. In this scheme, the ratio of k_{obs} for H1.5-Cys and H1.5 should correspond to the ratio of k_2 for a thiol attachment and k_2 for a non-thiol attachment (Equation (1)):

$$\frac{k_{obs,\ H1.5-Cys}}{k_{obs,\ H1.5}} \approx \frac{k_{2,\ thiol}}{k_{2,non-thiol}} \tag{1}$$

Moreover, this ratio should be roughly constant, and indeed it is (350 ± 80, 230 ± 60, and 320 ± 20 for 5K, 10K, and 30K, respectively). Our data therefore support a model where the peptides penetrate the PEG layer frequently, temporally sampling conformations near the AuNP-PEG interface, but adsorption only occurs occasionally when direct contact is made with the gold surface (enabling a more stable bond to occur).

2.3. Adsorption of Larger Proteins onto PEGylated AuNPs

To explore this scheme, we next tested the adsorption of proteins of varying sizes and thiol group content. The adsorption of the selected proteins is tight and in slow exchange timescale with no line broadening, rendering their NMR signals invisible upon attachment to the AuNP surface. Therefore, the remaining protein NMR signals correspond to the unbound protein amount [38,50]. Examples of quantifying protein unbound concentrations using 1D NMR for BSA and 2D NMR for K19C GB3 are presented in Figures S3 and S4, Supporting Information. As compared to the H1.5 peptide (1.4 kDa), WT GB3 has a size of 6.2 kDa. During the 72 h of our experiment, no WT GB3 was detected to adsorb at a statistically significant level onto 5K-PEG-AuNPs, and only a small amount of this protein penetrated into 10K- and 30K-PEG-AuNPs (Figure 3A,E). Specifically, 10% and 25% of the original 20 µM WT GB3 sample was adsorbed on the 10K and 30K PEG-SH molecules, respectively. For comparison, the H1.5 peptide reached ~60% adsorption after a similar incubation period on 30K-PEG-AuNPs. This suggests that, in the absence of high affinity thiol groups, small molecular size is critical for diffusion into the PEG layers, especially the denser PEG layer formed by smaller PEGs.

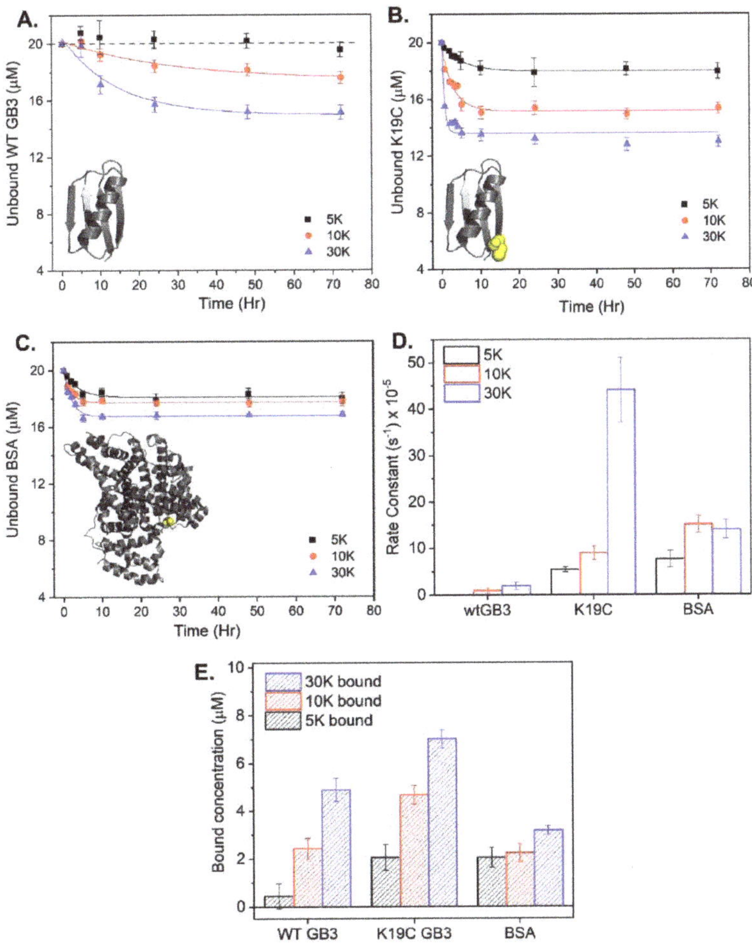

Figure 3. Adsorption of proteins onto PEGylated AuNPs. (**A**) Unbound WT GB3, (**B**) K19C GB3, and (**C**) BSA concentration to 5K, 10K, and 30K PEGylated AuNPs during the 0–72 h incubation period. The curves are the pseudo-first-order kinetics fit, and the error bars are the standard error of the mean. (**D**) Pseudo-first-order adsorption rate constants and (**E**) bound concentration for WT, K19C GB3, and BSA for different PEGylated AuNPs. Error bars in panels (**D**,**E**) represent uncertainties calculated from a weighted least squares minimization of the reduced chi-squared performed in OriginPro 2021b (Northampton, MA, USA).

In contrast, when one cysteine residue is introduced into GB3 (K19C GB3), both adsorption amount and adsorption rate increase significantly for all types of PEGylated AuNPs (Figure 3). For example, while the adsorption of WT GB3 on 5K-PEG-AuNPs was barely within the limits of detection for our experiment over 72 h, 2.1 ± 0.5 μM K19C GB3 is able to penetrate and bind onto these nanoparticles after only 10 h incubation. This difference becomes smaller as the PEG size is increased. While greater adsorption is always observed for K19C GB3, 10K-PEG-AuNPs allow nearly twice as much binding of K19C vs. WT GB3, but 30K-PEG-AuNPs allow only 1.4 times as much binding. In addition to increasing the final amount of adsorbed protein, introducing a cysteine residue also increases the rate at which adsorption occurs. We could not determine a pseudo-first order rate constant for WT GB3 adsorption to 5K-PEG-AuNPs, but rate constants could be reliably fit for the larger PEG sizes. Indeed, the calculated adsorption rate constants of K19C GB3 for both 10K-PEG-AuNP and 30K-PEG-AuNP are ~9 and ~23 times larger, respectively, than those measured for WT GB3 (Figure 3D).

In the context of Scheme 2, GB3 is apparently able to penetrate the PEG layer, and it can penetrate 10K and 30K PEG layers more efficiently than 5K PEG. Previously, it was theorized that adsorption isotherms tend to become independent of PEG length beyond 2000 Da (50 monomer units) for lysozyme and fibrinogen [51,52]. Both lysozyme (14 kDa) and fibrinogen (300 kDa) are much larger than GB3 (6.2 kDa), and for GB3 we see a size dependent effect. In addition, taking the ratio k_{obs} for K19C GB3 and k_{obs} for WT GB3 does not appear to produce a constant value as it did with H1.5 (9 ± 4 for 10K vs. 23 ± 10 for 30K), suggesting that the complexity of larger proteins complicates Scheme 2. This may be as the size of GB3 is similar to the size of the PEG molecules themselves, and reorientation in the PEG layer becomes more difficult when the protein size is similar to the passivating PEG. As the PEG size increases, the attachment density decreases (Table 1), and GB3 is more easily able to slip through the passivating layer. Thus, the size and shape of the protein relative to the PEG density will likely influence whether the first step in scheme two is fast-to-equilibrium. Additional work is needed to explore this hypothesis; nevertheless, these results highlight the importance of free and surface cysteine (thiol) groups in protein adsorption onto PEGylated AuNP. The high affinity of thiol towards gold not only accelerates the adsorption rate of large biomolecules, but also increases the final bound amount.

On the other hand, BSA has a size ~10 times that of the GB3 protein. With the same number of free surface thiol groups as K19C GB3, the size effect of protein adsorption on PEGylated AuNP can be examined. Previous experiments using fluorescence spectroscopy found that BSA was buried inside the 10K PEG layer on an AuNP [14,26]. Our studies indicate that, due to the size of BSA, only a small concentration of BSA is adsorbed by all three types of PEGylated AuNP surfaces. The maximum adsorption is again observed for the 30K-PEG-AuNPs. The same trend that larger PEG promotes greater adsorption is observed, although the difference between 5K PEG and 30K PEG is markedly less for BSA vs. K19C GB3. Interestingly, even large proteins are able to penetrate relatively dense monolayers of PEG on an AuNP surface. While the amount of adsorbed protein is small, the value is reproducible, and such a small number may be able to alter the immune response to an engineered nanoparticle, especially if those nanoparticles experience long circulation times.

For comparison, the binding of these biomolecules onto non-PEGylated AuNPs is extremely fast and results in higher binding capacities. Previous studies show that ~90% of protein binding to citrate-coated AuNPs occurs in the initial ~5 min, and adsorption is completed within an hour [38,39]. The binding capacities (molecules per AuNp) for GSH, WT GB3 and BSA on bare AuNPs were determined to be 1430 ± 90, 177 ± 20, and 30 ± 10, respectively [37]. In contrast, it takes 5–20 h for GSH, WT GB3, and BSA to reach their maximum binding capacities (molecules per AuNP) of 123 ± 2, 41 ± 4, and 123 ± 2, 26 ± 2, respectively on AuNP-PEG-30K. The binding rates and final amounts are even lower on AuNP-PEG-5K and AuNP-PEG-10K.

3. Discussion

The observation that proteins can penetrate PEG-passivated surfaces is not new [29,30]; however, systematic studies to identify which protein properties favor fouling of passivated AuNPs can be useful in improving strategies for limiting protein binding. Here, we design a simple test that examines how protein size, charge, and surface cysteine content influences adsorption for several different PEGylation densities on AuNPs, holding the size of the core AuNP constant. Our approach employs a newly developed external NMR referencing method for quantifying protein binding to AuNPs in situ, and our observations do not require displacement of proteins or treatment of the AuNPs [38]. This approach enables straightforward measurements of adsorption kinetics, provided no line broadening is observed in the biomolecular NMR spectra. We observe that both the adsorption rate and the number of biomolecules adsorbed onto PEGylated AuNPs increase as the size of PEG-SH increases, from 5K-, 10K-, to 30K-PEG-AuNPs. This validates previous observations that smaller PEGs have a better passivation effect on AuNPs due to their higher grafting density, making diffusion of biomolecules into the dense PEG polymer chains less efficient, and leaving less AuNP surface for biomolecule loading [16,34,51–53]. On the contrary, larger PEG can stabilize AuNPs as well as allow biomolecules to diffuse efficiently and bind to the AuNPs. With all peptides and proteins tested in this work, maximum binding for 30K-PEG-AuNP is attained within several hours. These results should provide insights into how to select PEG size for different applications of PEGylated AuNPs.

Comparing the adsorption behaviors of H1.5, GSH, and WT GB3 onto PEGylated AuNPs, we conclude that molecules with small sizes (significantly smaller than the PEG size) can efficiently penetrate the PEG layers, rapidly sampling conformations near the surface as shown in Scheme 2. The presence of thiol groups for small peptides can dramatically enhance adsorption. Once the protein becomes larger than the PEG itself, however, the situation is more complex and likely depends on the protein itself. Surface Cys residues are still important, as demonstrated by the difference between WT and K19C GB3. Steric hindrance from the densely populated PEG chains is effective at reducing protein binding, but the introduction of only one cysteine residue allows even weakly associated proteins to adsorb to the AuNP surface. Even BSA, which is approximately 10 times the size of WT GB3, can be bound to 5K-PEG-AuNP, the PEGylated AuNP with highest grafting density. The survey of these large biomolecules demonstrates high affinity groups are required for large molecules to be loaded onto PEGylated AuNPs.

In conclusion, our study indicates that larger PEG molecules are less effective in passivating biomolecules molecules with small sizes (GSH and H1.5) and macromolecules with thiol groups (H1.5-Cys, K19C GB3, and BSA). Larger PEG constructs adopt a "mushroom"-like structure, leaving voids in between the PEG chains [16]. Due to this mushroom-like structure, the surface density of larger PEG on AuNPs is lower than the short PEG chains. This phenomenon allows the proteins to easily penetrate the PEG layer. Shorter PEG chains are much more effective in passivating the AuNP surface from biomolecules, but the presence of surface cysteine residues appears to nullify the passivating effect, especially at timescales greater than 12 h, which are relevant for the pharmacokinetics of AuNP-based therapeutics.

4. Materials and Methods

4.1. Synthesis of Citrate-Stabilized Gold Nanoparticles

Spherical 15 nm AuNPs were synthesized via citric acid reduction using principles of the Turkevic synthesis method [54,55]. Tetrachloroauric Acid ($HAuCl_4$) and sodium citrate dihydrate were purchased from Sigma-Aldrich. After 100 mL of 0.3 mM $HAuCl_4$ had been heated to boiling, 2 mL of 34 mM sodium citrate solution was immediately mixed with the gold solution. This mixture was stirred with heating for an additional 20 min before being cooled to room temperature. The cooled solution was then centrifuged for 45 min at $9000\times g$. The concentrated AuNP sample was then extracted and sonicated for 6 min (in 1-min intervals) at a power level of 1 on a Branson sonicator. The sonicated sample was characterized via UV-visible spectroscopy, dynamic light scattering, and transmission electron microscopy for size and conformity [37]. For 15 nm AuNPs, it was expected that the maximum absorbance should be at 520 nm with an extinction coefficient of 3.9×10^8 M^{-1} cm^{-1} [56–59].

4.2. Protein Preparation

^{15}N-labeled WT GB3 and its variant K19C GB3 were expressed and purified according to previously published methods [37,59,60]. Protein purity was established using SDS-PAGE electrophoresis. The H1.5 peptide (KVAKpSPKKAKAW, where pS represents phosphoserine) [61] and H1.5-Cys (KVAKpSPKKCKAW) were purchased from GenScript (Piscataway, NJ, USA) and BioMatik (Wilmington, DE, USA) at 95% purity and used after dissolving in appropriate buffer. GSH and BSA were purchased from Sigma Aldrich and were used without any additional purification. The concentration of GSH was determined using the molar mass; concentrations for all other peptides and proteins were determined using the calculated extinction coefficient at 280 nm [62].

4.3. PEGylated Gold Nanoparticle Preparation

The thiolated 5K, 10K, and 30K poly(ethylene glycol) monomethyl ether compounds (PEG-SH) were purchased from Laysan Bio, Inc. (Arab, AL, USA). The PEG-SH compounds were dissolved in MilliQ water containing 50 mM TCEP to maintain reducing conditions. A solution of 120 nM AuNPs was incubated with 100 μM of the three different sizes of PEG to fully saturate the AuNP surface. After the overnight incubation at room temperature, the solutions were centrifuged at $9000 \times g$ and washed three times with buffer to remove any unbound PEG molecules. Afterward, PEG-coated AuNPs were resuspended in buffer. The PEGylated AuNPs were characterized using UV-Vis spectroscopy and dynamic light scattering to confirm the functionalization. The same extinction coefficient was used for PEGylated AuNPs as for bare AuNPs. Atomic absorption spectroscopy confirmed that this approximation was accurate to within 6% error (Table S2). The PEG density on the AuNP surface was determined as follows. A total of 120 μM thiolated PEG was mixed with 120 nM 15-nm AuNPs. After 1-h incubation, all PEG bound AuNPs were thoroughly spun down (10 min at $21,000 \times g$) and the bound fraction of PEG was determined by the reduction in thiol concentration in the supernatant using Ellman's reagent (5,5′-dithio-bis (2-nitrobenzoic acid)) [63]. The coordination efficiency (θ) of PEG was calculated using $\theta = 1 - [SH]_{free}/[SH]_{total}$, where $[SH]_{total}$ and $[SH]_{free}$ correspond to concentration of thiolated PEG in the supernatant before and after binding to AuNP. The PEG grafting density (ρ_{PEG}) was calculated using formula $\rho_{PEG} = \theta\, \rho_{graft}$, where ρ_{graft} is the number of initial PEG molecules available per nm^2 AuNP surface.

4.4. Transmission Electron Miscroscopy (TEM) Measurement of PEGylated AuNPs

Aliquots of 5 μL of 2 nM PEGylated AuNP solution was deposited on Formvar-coated copper grids. The excess liquid was wicked away, and the remaining thin film on the grid was allowed to dry. Prepared grids were imaged using JEOL 2100 with an accelerating voltage of 200 kV. TEM was performed at the Institute for Imaging and Analytical Technologies (I^{2}AT) at Mississippi State University.

4.5. NMR Adsorption Measurements

For the control sample, 20 μM protein or peptide was prepared with 20 mM phosphate buffer at pH 6.5. Protein samples were mixed with AuNPs at concentration of 120 nM AuNPs. Quantitation was performed using an external standard, described previously [38]. This approach is effective when adsorption is slow on the NMR timescale, and when no line broadening is observed, as occurs here [38,50]. A solution of 50 mM TCEP and pH 6.5 PIPES buffer was used in the sample preparation of proteins and peptides containing thiol groups (GSH, H1.5-Cys, and K19C GB3). The samples were incubated for 6–72 hr before taking the 1D ^{1}H NMR spectra for GSH, H1.5, and BSA. ^{1}H-^{15}N HSQC spectra were collected to measure the bound protein concentration of WT GB3 and K19C GB3. The NMR spectra were recorded at 25 °C using a 600 MHz Bruker Avance III cryoprobe-equipped NMR spectrometer. NMR spectra were processed using TOPSPIN 3.1 software. The bound protein concentration was measured by using the DSS peak as an external reference and integrating the protein amide signal with and without AuNPs at different time intervals [37,38]. First, all spectra were normalized to the external DSS reference peak. Then, the ratio of each peak intensity relative to the protein signal in the absence of nanoparticles was used to quantify the amount of protein remaining in solution. This ratio (r) represents the quantitative loss in protein signal, and in a standard 2D HSQC NMR spectrum can be averaged over all non-overlapping protein peaks. The concentration of free protein is then calculated as $(1 - r)C_0$ where C_0 is the initial concentration of protein in the absence of AuNPs (here, C_0 is 20 μM). Additional details and experimental considerations are discussed in Xu et al. [38]. For 1D measurements (GSH, H1.5, H1.5-Cys, and BSA), the amide region of a water suppression experiment is used to calculate r for the entire amide proton region, as described in Wang et al. [37]. Examples of the 1D and HSQC NMR spectra are provided in the Supporting Information as Figures S3 and S4. Error bars are calculated as the standard error of the mean from at least three independent measurements.

4.6. Dynamic Light Scattering

A Wyatt DynaPro NanoStar DLS instrument was used to measure the NP size distributions. After equilibration for 1 h at room temperature, the solution was diluted 5-fold before transferring to a disposable microcuvette for measurement. The hydrodynamic diameters of the AuNPs were measured using the regularization fit functionality of the DYNAMICS software. For each measurement (with or without PEG), the average value of three independently prepared samples is reported and the uncertainty is calculated as the standard error of the mean. Zeta potential measurements were performed on an Anton Paar Litesizer 500 instrument using Kalliope software.

Supplementary Materials: The following are available online. Figure S1: UV-vis spectra of PEGylated AuNPs; Figure S2: Transmission electron microscopy (TEM) characterization of PEG-grafted gold nanoparticles; Figure S3: Example of quantifying protein unbound concentrations using 1D NMR for BSA; Figure S4: Example of quantifying protein bound concentrations using 2D NMR for K19C GB3; Table S1: Summary of bound concentration ([bound]) of ligand when mixing 20 μM ligand with 120 nM 15-nm AuNPs and observed rate constants (k_{obs}) of 20 μM different ligands onto 120 nM PEGylated AuNPs used in this work; Table S2: Comparison of AuNP-PEG concentrations determined by atomic absorption spectroscopy (AAS) and concentrations determined using the extinction coefficient at 520 nm (3.9×10^8 M^{-1} cm^{-1}), as described in the text.

Author Contributions: Conceptualization, methodology and analysis: Y.R.P., J.X.X., D.L.A. and N.C.F. Investigation: Y.R.P., J.X.X., D.L.A., I.A. and A.C.H. Writing—original draft: Y.R.P., J.X.X. and N.C.F. Writing—review and editing: J.X.X. and N.C.F. All authors have read and agreed to the published version of the manuscript.

Funding: This work was supported by the National Institute of Allergy and Infectious Diseases of the National Institutes of Health under grant number R01AI139479 and the National Science Foundation under grants 1818090 and 1852527.

Institutional Review Board Statement: Not applicable.

Informed Consent Statement: Not applicable.

Data Availability Statement: All data from this study are available in the figures and Supporting Information. Tabulated data points are available from the authors upon request.

Acknowledgments: We thank Rahul Yadav and Becca Hill for careful reading of the manuscript and for thoughtful comments and suggestions.

Conflicts of Interest: The authors declare no conflict of interest.

Sample Availability: Protein expression vectors for GB3 are available from the authors upon request. All other proteins and peptides are available commercially.

References

1. Jokerst, J.V.; Lobovkina, T.; Zare, R.N.; Gambhir, S.S. Nanoparticle PEGylation for imaging and therapy. *Nanomedicine* **2011**, *6*, 715–728. [CrossRef]
2. Sanchez-Cano, C.; Carril, M. Recent Developments in the Design of Non-Biofouling Coatings for Nanoparticles and Surfaces. *Int. J. Mol. Sci.* **2020**, *21*, E1007. [CrossRef]
3. Behzadi, S.; Serpooshan, V.; Tao, W.; Hamaly, M.A.; Alkawareek, M.Y.; Dreaden, E.C.; Brown, D.; Alkilany, A.M.; Farokhzad, O.C.; Mahmoudi, M. Cellular uptake of nanoparticles: Journey inside the cell. *Chem. Soc. Rev.* **2017**, *46*, 4218–4244. [CrossRef] [PubMed]
4. Guerrini, L.; Alvarez-Puebla, R.A.; Pazos-Perez, N. Surface Modifications of Nanoparticles for Stability in Biological Fluids. *Materials* **2018**, *11*, 1154. [CrossRef] [PubMed]
5. Miller, H.A.; Magsam, A.W.; Tarudji, A.W.; Romanova, S.; Weber, L.; Gee, C.C.; Madsen, G.L.; Bronich, T.K.; Kievit, F.M. Evaluating differential nanoparticle accumulation and retention kinetics in a mouse model of traumatic brain injury via Ktrans mapping with MRI. *Sci. Rep.* **2019**, *9*, 16099. [CrossRef] [PubMed]
6. Gustafson, H.H.; Holt-Casper, D.; Grainger, D.W.; Ghandehari, H. Nanoparticle Uptake: The Phagocyte Problem. *Nano Today* **2015**, *10*, 487–510. [CrossRef]
7. Hoshyar, N.; Gray, S.; Han, H.; Bao, G. The effect of nanoparticle size on in vivo pharmacokinetics and cellular interaction. *Nanomedicine* **2016**, *11*, 673–692. [CrossRef]
8. Zhang, X.-D.; Wu, D.; Shen, X.; Liu, P.-X.; Yang, N.; Zhao, B.; Zhang, H.; Sun, Y.-M.; Zhang, L.-A.; Fan, F.-Y. Size-dependent in vivo toxicity of PEG-coated gold nanoparticles. *Int. J. Nanomed.* **2011**, *6*, 2071–2081. [CrossRef] [PubMed]
9. Bazile, D.; Prud'homme, C.; Bassoullet, M.-T.; Marlard, M.; Spenlehauer, G.; Veillard, M. Stealth Me.PEG-PLA nanoparticles avoid uptake by the mononuclear phagocytes system. *J. Pharm. Sci.* **1995**, *84*, 493–498. [CrossRef]
10. Cho, W.-S.; Cho, M.; Jeong, J.; Choi, M.; Cho, H.-Y.; Han, B.S.; Kim, S.H.; Kim, H.O.; Lim, Y.T.; Chung, B.H.; et al. Acute toxicity and pharmacokinetics of 13 nm-sized PEG-coated gold nanoparticles. *Toxicol. Appl. Pharmacol.* **2009**, *236*, 16–24. [CrossRef] [PubMed]
11. Li, D.; Wang, F.; Di, H.; Liu, X.; Zhang, P.; Zhou, W.; Liu, D. Cross-Linked Poly(ethylene glycol) Shells for Nanoparticles: Enhanced Stealth Effect and Colloidal Stability. *Langmuir* **2019**, *35*, 8799–8805. [CrossRef] [PubMed]
12. Qie, Y.; Yuan, H.; von Roemeling, C.A.; Chen, Y.; Liu, X.; Shih, K.D.; Knight, J.A.; Tun, H.W.; Wharen, R.E.; Jiang, W.; et al. Surface modification of nanoparticles enables selective evasion of phagocytic clearance by distinct macrophage phenotypes. *Sci. Rep.* **2016**, *6*, 26269. [CrossRef] [PubMed]
13. Kuhn, D.A.; Vanhecke, D.; Michen, B.; Blank, F.; Gehr, P.; Petri-Fink, A.; Rothen-Rutishauser, B. Different endocytotic uptake mechanisms for nanoparticles in epithelial cells and macrophages. *Beilstein J. Nanotechnol.* **2014**, *5*, 1625–1636. [CrossRef]
14. Pelaz, B.; del Pino, P.; Maffre, P.; Hartmann, R.; Gallego, M.; Rivera-Fernández, S.; de la Fuente, J.M.; Nienhaus, G.U.; Parak, W.J. Surface Functionalization of Nanoparticles with Polyethylene Glycol: Effects on Protein Adsorption and Cellular Uptake. *ACS Nano* **2015**, *9*, 6996–7008. [CrossRef] [PubMed]
15. Cho, W.-S.; Cho, M.; Jeong, J.; Choi, M.; Han, B.S.; Shin, H.-S.; Hong, J.; Chung, B.H.; Jeong, J.; Cho, M.-H. Size-dependent tissue kinetics of PEG-coated gold nanoparticles. *Toxicol. Appl. Pharmacol.* **2010**, *245*, 116–123. [CrossRef] [PubMed]
16. Li, M.; Jiang, S.; Simon, J.; Paßlick, D.; Frey, M.-L.; Wagner, M.; Mailänder, V.; Crespy, D.; Landfester, K. Brush Conformation of Polyethylene Glycol Determines the Stealth Effect of Nanocarriers in the Low Protein Adsorption Regime. *Nano Lett.* **2021**, *21*, 1591–1598. [CrossRef]
17. García, I.; Sánchez-Iglesias, A.; Henriksen-Lacey, M.; Grzelczak, M.; Penadés, S.; Liz-Marzán, L.M. Glycans as Biofunctional Ligands for Gold Nanorods: Stability and Targeting in Protein-Rich Media. *J. Am. Chem. Soc.* **2015**, *137*, 3686–3692. [CrossRef]
18. Dai, Q.; Walkey, C.; Chan, W.C.W. Polyethylene Glycol Backfilling Mitigates the Negative Impact of the Protein Corona on Nanoparticle Cell Targeting. *Angew. Chem. Int. Ed.* **2014**, *53*, 5093–5096. [CrossRef] [PubMed]
19. Blanco, E.; Shen, H.; Ferrari, M. Principles of nanoparticle design for overcoming biological barriers to drug delivery. *Nat. Biotechnol.* **2015**, *33*, 941–951. [CrossRef]
20. Niidome, T.; Yamagata, M.; Okamoto, Y.; Akiyama, Y.; Takahashi, H.; Kawano, T.; Katayama, Y.; Niidome, Y. PEG-modified gold nanorods with a stealth character for in vivo applications. *J. Controlled Release* **2006**, *114*, 343–347. [CrossRef] [PubMed]
21. Perry, J.L.; Reuter, K.G.; Kai, M.P.; Herlihy, K.P.; Jones, S.W.; Luft, J.C.; Napier, M.; Bear, J.E.; DeSimone, J.M. PEGylated PRINT Nanoparticles: The Impact of PEG Density on Protein Binding, Macrophage Association, Biodistribution, and Pharmacokinetics. *Nano Lett.* **2012**, *12*, 5304–5310. [CrossRef]
22. Suk, J.S.; Xu, Q.; Kim, N.; Hanes, J.; Ensign, L.M. PEGylation as a strategy for improving nanoparticle-based drug and gene delivery. *Adv. Drug Del. Rev.* **2016**, *99*, 28–51. [CrossRef]
23. Li Volsi, A.; Jimenez de Aberasturi, D.; Henriksen-Lacey, M.; Giammona, G.; Licciardi, M.; Liz-Marzán, L.M. Inulin coated plasmonic gold nanoparticles as a tumor-selective tool for cancer therapy. *J. Mater. Chem. B* **2016**, *4*, 1150–1155. [CrossRef] [PubMed]

24. Oyewumi, M.O.; Yokel, R.A.; Jay, M.; Coakley, T.; Mumper, R.J. Comparison of cell uptake, biodistribution and tumor retention of folate-coated and PEG-coated gadolinium nanoparticles in tumor-bearing mice. *J. Control. Release* **2004**, *95*, 613–626. [CrossRef] [PubMed]

25. Spadavecchia, J.; Movia, D.; Moore, C.; Maguire, C.M.; Moustaoui, H.; Casale, S.; Volkov, Y.; Prina-Mello, A. Targeted polyethylene glycol gold nanoparticles for the treatment of pancreatic cancer: From synthesis to proof-of-concept in vitro studies. *Int. J. Nanomed.* **2016**, *11*, 791–822. [CrossRef] [PubMed]

26. Polo, E.; Araban, V.; Pelaz, B.; Alvarez, A.; Taboada, P.; Mahmoudi, M.; del Pino, P. Photothermal effects on protein adsorption dynamics of PEGylated gold nanorods. *Appl. Mater. Today* **2019**, *15*, 599–604. [CrossRef]

27. Zhang, M.; Li, X.H.; Gong, Y.D.; Zhao, N.M.; Zhang, X.F. Properties and biocompatibility of chitosan films modified by blending with PEG. *Biomaterials* **2002**, *23*, 2641–2648. [CrossRef]

28. Zalipsky, S. Chemistry of polyethylene glycol conjugates with biologically active molecules. *Adv. Drug Del. Rev.* **1995**, *16*, 157–182. [CrossRef]

29. Gref, R.; Lück, M.; Quellec, P.; Marchand, M.; Dellacherie, E.; Harnisch, S.; Blunk, T.; Müller, R.H. 'Stealth' corona-core nanoparticles surface modified by polyethylene glycol (PEG): Influences of the corona (PEG chain length and surface density) and of the core composition on phagocytic uptake and plasma protein adsorption. *Colloids Surf. B. Biointerfaces* **2000**, *18*, 301–313. [CrossRef]

30. Du, Y.; Jin, J.; Liang, H.; Jiang, W. Structural and Physicochemical Properties and Biocompatibility of Linear and Looped Polymer-Capped Gold Nanoparticles. *Langmuir* **2019**, *35*, 8316–8324. [CrossRef]

31. Hristov, D.R.; Lopez, H.; Ortin, Y.; O'Sullivan, K.; Dawson, K.A.; Brougham, D.F. Impact of dynamic sub-populations within grafted chains on the protein binding and colloidal stability of PEGylated nanoparticles. *Nanoscale* **2021**, *13*, 5344–5355. [CrossRef]

32. Partikel, K.; Korte, R.; Stein, N.C.; Mulac, D.; Herrmann, F.C.; Humpf, H.-U.; Langer, K. Effect of nanoparticle size and PEGylation on the protein corona of PLGA nanoparticles. *Eur. J. Pharm. Biopharm.* **2019**, *141*, 70–80. [CrossRef] [PubMed]

33. Blume, J.E.; Manning, W.C.; Troiano, G.; Hornburg, D.; Figa, M.; Hesterberg, L.; Platt, T.L.; Zhao, X.; Cuaresma, R.A.; Everley, P.A.; et al. Rapid, deep and precise profiling of the plasma proteome with multi-nanoparticle protein corona. *Nat. Commun.* **2020**, *11*, 3662. [CrossRef]

34. Walkey, C.D.; Olsen, J.B.; Guo, H.; Emili, A.; Chan, W.C.W. Nanoparticle Size and Surface Chemistry Determine Serum Protein Adsorption and Macrophage Uptake. *J. Am. Chem. Soc.* **2012**, *134*, 2139–2147. [CrossRef] [PubMed]

35. Zhou, H.; Fan, Z.; Li, P.Y.; Deng, J.; Arhontoulis, D.C.; Li, C.Y.; Bowne, W.B.; Cheng, H. Dense and Dynamic Polyethylene Glycol Shells Cloak Nanoparticles from Uptake by Liver Endothelial Cells for Long Blood Circulation. *ACS Nano* **2018**, *12*, 10130–10141. [CrossRef] [PubMed]

36. Li, S.-D.; Huang, L. Pharmacokinetics and Biodistribution of Nanoparticles. *Mol. Pharm.* **2008**, *5*, 496–504. [CrossRef] [PubMed]

37. Wang, A.; Vangala, K.; Vo, T.; Zhang, D.; Fitzkee, N.C. A Three-Step Model for Protein–Gold Nanoparticle Adsorption. *J. Phys. Chem. C* **2014**, *118*, 8134–8142. [CrossRef]

38. Xu, J.X.; Alom, M.S.; Fitzkee, N.C. Quantitative Measurement of Multiprotein Nanoparticle Interactions Using NMR Spectroscopy. *Anal. Chem.* **2021**, *93*, 11982–11990. [CrossRef]

39. Xu, J.X.; Fitzkee, N.C. Solution NMR of Nanoparticles in Serum: Protein Competition Influences Binding Thermodynamics and Kinetics. *Front. Physiol.* **2021**, *12*, 715419. [CrossRef]

40. Siriwardana, K.; Gadogbe, M.; Ansar, S.M.; Vasquez, E.S.; Collier, W.E.; Zou, S.; Walters, K.B.; Zhang, D. Ligand Adsorption and Exchange on Pegylated Gold Nanoparticles. *J. Phys. Chem. C* **2014**, *118*, 11111–11119. [CrossRef]

41. Tsai, D.-H.; Lu, Y.-F.; DelRio, F.W.; Cho, T.J.; Guha, S.; Zachariah, M.R.; Zhang, F.; Allen, A.; Hackley, V.A. Orthogonal analysis of functional gold nanoparticles for biomedical applications. *Anal. Bioanal. Chem.* **2015**, *407*, 8411–8422. [CrossRef]

42. Xue, Y.; Li, X.; Li, H.; Zhang, W. Quantifying thiol–gold interactions towards the efficient strength control. *Nat. Commun.* **2014**, *5*, 4348. [CrossRef] [PubMed]

43. Inkpen, M.S.; Liu, Z.F.; Li, H.; Campos, L.M.; Neaton, J.B.; Venkataraman, L. Non-chemisorbed gold–sulfur binding prevails in self-assembled monolayers. *Nat. Chem.* **2019**, *11*, 351–358. [CrossRef]

44. Rahme, K.; Chen, L.; Hobbs, R.G.; Morris, M.A.; O'Driscoll, C.; Holmes, J.D. PEGylated gold nanoparticles: Polymer quantification as a function of PEG lengths and nanoparticle dimensions. *RSC Adv.* **2013**, *3*, 6085–6094. [CrossRef]

45. Leff, D.V.; Brandt, L.; Heath, J.R. Synthesis and Characterization of Hydrophobic, Organically-Soluble Gold Nanocrystals Functionalized with Primary Amines. *Langmuir* **1996**, *12*, 4723–4730. [CrossRef]

46. Selvakannan, P.R.; Mandal, S.; Phadtare, S.; Pasricha, R.; Sastry, M. Capping of Gold Nanoparticles by the Amino Acid Lysine Renders Them Water-Dispersible. *Langmuir* **2003**, *19*, 3545–3549. [CrossRef]

47. Jana, N.R.; Gearheart, L.; Murphy, C.J. Wet chemical synthesis of silver nanorods and nanowires of controllable aspect ratio. *Chem. Commun.* **2001**, 617–618. [CrossRef]

48. Provorse, M.R.; Aikens, C.M. Binding of carboxylates to gold nanoparticles: A theoretical study of the adsorption of formate on Au20. *Comput. Theor. Chem.* **2012**, *987*, 16–21. [CrossRef]

49. Krishna, M.M.G.; Hoang, L.; Lin, Y.; Englander, S.W. Hydrogen exchange methods to study protein folding. *Methods* **2004**, *34*, 51–64. [CrossRef]

50. Perera, Y.R.; Hill, R.A.; Fitzkee, N.C. Protein Interactions with Nanoparticle Surfaces: Highlighting Solution NMR Techniques. *Isr. J. Chem.* **2019**, *59*, 962–979. [CrossRef] [PubMed]

51. Szleifer, I. Protein Adsorption on Surfaces with Grafted Polymers: A Theoretical Approach. *Biophys. J.* **1997**, *72*, 595–612. [CrossRef]
52. Unsworth, L.D.; Sheardown, H.; Brash, J.L. Protein-Resistant Poly(ethylene oxide)-Grafted Surfaces: Chain Density-Dependent Multiple Mechanisms of Action. *Langmuir* **2008**, *24*, 1924–1929. [CrossRef]
53. Stanicki, D.; Larbanoix, L.; Boutry, S.; Vangijzegem, T.; Ternad, I.; Garifo, S.; Muller, R.N.; Laurent, S. Impact of the chain length on the biodistribution profiles of PEGylated iron oxide nanoparticles: A multimodal imaging study. *J. Mater. Chem. B* **2021**, *9*, 5055–5068. [CrossRef] [PubMed]
54. Turkevich, J.; Stevenson, P.C.; Hillier, J. A study of the nucleation and growth processes in the synthesis of colloidal gold. *Discuss. Faraday Soc.* **1951**, *11*, 55–75. [CrossRef]
55. Frens, G. Controlled Nucleation for the Regulation of the Particle Size in Monodisperse Gold Suspensions. *Nat. Phys. Sci.* **1973**, *241*, 20–22. [CrossRef]
56. Link, S.; El-Sayed, M.A. Size and Temperature Dependence of the Plasmon Absorption of Colloidal Gold Nanoparticles. *J. Phys. Chem. B* **1999**, *103*, 4212–4217. [CrossRef]
57. Jain, P.K.; Lee, K.S.; El-Sayed, I.H.; El-Sayed, M.A. Calculated Absorption and Scattering Properties of Gold Nanoparticles of Different Size, Shape, and Composition: Applications in Biological Imaging and Biomedicine. *J. Phys. Chem. B* **2006**, *110*, 7238–7248. [CrossRef] [PubMed]
58. Liu, X.; Atwater, M.; Wang, J.; Huo, Q. Extinction coefficient of gold nanoparticles with different sizes and different capping ligands. *Colloids Surf. B. Biointerfaces* **2007**, *58*, 3–7. [CrossRef]
59. Woods, K.E.; Perera, Y.R.; Davidson, M.B.; Wilks, C.A.; Yadav, D.K.; Fitzkee, N.C. Understanding Protein Structure Deformation on the Surface of Gold Nanoparticles of Varying Size. *J. Phys. Chem. C* **2016**, *120*, 27944–27953. [CrossRef]
60. Wang, A.; Perera, Y.R.; Davidson, M.B.; Fitzkee, N.C. Electrostatic Interactions and Protein Competition Reveal a Dynamic Surface in Gold Nanoparticle–Protein Adsorption. *J. Phys. Chem. C* **2016**, *120*, 24231–24239. [CrossRef]
61. Jinasena, D.; Simmons, R.; Gyamfi, H.; Fitzkee, N.C. Molecular Mechanism of the Pin1–Histone H1 Interaction. *Biochemistry* **2019**, *58*, 788–798. [CrossRef] [PubMed]
62. Pace, C.N.; Vajdos, F.; Fee, L.; Grimsley, G.; Gray, T. How to measure and predict the molar absorption coefficient of a protein. *Protein Sci.* **1995**, *4*, 2411–2423. [CrossRef] [PubMed]
63. Ellman, G.L. Tissue sulfhydryl groups. *Arch. Biochem. Biophys.* **1959**, *82*, 70–77. [CrossRef]

Review

Alpha-Synuclein—Nanoparticle Interactions: Understanding, Controlling and Exploiting Conformational Plasticity

Mariapina D'Onofrio , **Francesca Munari and Michael Assfalg** *

Department of Biotechnology, University of Verona, 37134 Verona, Italy; mariapina.donofrio@univr.it (M.D.); francesca.munari@univr.it (F.M.)
* Correspondence: michael.assfalg@univr.it; Tel.: +39-045-8027939

Academic Editor: Pil Seok Chae
Received: 25 September 2020; Accepted: 27 November 2020; Published: 29 November 2020

Abstract: Alpha-synuclein (αS) is an extensively studied protein due to its involvement in a group of neurodegenerative disorders, including Parkinson's disease, and its documented ability to undergo aberrant self-aggregation resulting in the formation of amyloid-like fibrils. In dilute solution, the protein is intrinsically disordered but can adopt multiple alternative conformations under given conditions, such as upon adsorption to nanoscale surfaces. The study of αS-nanoparticle interactions allows us to better understand the behavior of the protein and provides the basis for developing systems capable of mitigating the formation of toxic aggregates as well as for designing hybrid nanomaterials with novel functionalities for applications in various research areas. In this review, we summarize current progress on αS-nanoparticle interactions with an emphasis on the conformational plasticity of the biomolecule.

Keywords: alpha-synuclein; amyloid fibrils; conformational flexibility; protein adsorption; protein aggregation; nano-bio interface; nanocomposite; nanoparticles; supramolecular assembly

---

1. Introduction

Alpha-synuclein (αS) is a paradigmatic and one of the most extensively investigated intrinsically disordered proteins (IDPs) [1,2]. It is an abundant neuronal protein which localizes predominantly to presynaptic terminals and binds to small synaptic vesicles [3]. The biological function of the protein remains enigmatic, although increasing evidence supports its participation in neurotransmission and synaptic plasticity, including roles in synaptic vesicle recycling and neurotransmitter synthesis and release [4–6]. αS has also been reported to interact with and affect a variety of proteins [7]. Soon after its discovery, the protein became infamous for being strongly linked, genetically and pathologically, to Parkinson's disease (PD) and other neurodegenerative diseases characterized by abnormal accumulation of insoluble αS deposits [8–10].

Encoded by the SNCA gene, αS is a 14-kDa polypeptide which can be broadly divided into three domains: an amphiphilic N-terminal region (residues 1–60) that contains four imperfect eleven-residue amino acid repeats, a hydrophobic, amyloidogenic domain (residues 61–95, referred to as non-amyloid-β component, NAC, domain) with three additional repeats, and an acidic C terminus (residues 96–140) [9,11]. In dilute solution, αS is unstructured and best described as a dynamic ensemble of interconverting conformations [12–14]. Mounting evidence indicates that native oligomers exist in cells, which exhibit greater aggregation resistance than the disordered monomeric species [15–18]. Environmental changes, binding events and other stimuli may promote the transition from the soluble monomeric and small oligomeric states to higher-level oligomers, fibrils (highly ordered supramolecular nanostructures) and amorphous aggregates [12,15,19,20]. The high protein solubility and the possibility to trigger conformational changes in vitro upon exposure to specific environments and stimuli, have made αS a popular model for structural and aggregation studies of IDPs [1,21].

In the context of amyloidogenic, unfolded peptides and proteins like αS, nanoparticles (NPs) have attracted interest as artificial receptors or chaperones against the formation of toxic aggregates [22]. NP surfaces, acting as a scaffold for protein adsorption, could provide the means to redirect aggregation pathways, sequester/correct misfolded structures, retard or accelerate aggregation and disaggregate assemblies. NPs are versatile platforms, they can be prepared in a wide range of sizes and with diverse surface chemistries. In principle, by careful design of the NPs it should be possible to control their interactions with biological components and develop artificial receptors capable of biomolecular recognition [23–28]. Among the vast array of potential applications that rely on optimized target recognition, NPs represent a promising alternative to conventional small drugs for targeting protein-protein interactions associated with pathological conditions [24] and indeed have emerged as a new class of therapeutics [29].

The nanometer scale of their size makes NPs able to interact with cellular systems and biomolecular networks and to reach targets of biomedical interest [30]. Upon exposure to biological media, NPs tend to be covered by a layer of biomolecules, generally consisting of an internal, long-lived layer (termed hard corona) and a more loosely associated, external layer (termed soft corona) [30]. The protein corona mediates the interactions with the living systems and determines the physiological responses [30,31]. Unintended protein adsorption onto NPs may perturb the protein's activity as a consequence of binding-induced changes in structure, stability or the exposure of recognition sites [32]. It is thus essential to characterize both the structural and dynamic organization of the corona and the modes of binding of distinct biomolecules to the NP surface [33].

The ability of ad-hoc prepared NPs to target and associate to specific proteins can be extended to the development of novel hybrid materials composed of proteins and NPs, which feature unique attributes not attainable with the separate components [34]. Protein-NP bioconjugation is an attractive method to fabricate functional materials enabling applications in sensing, delivery and other nanotechnological areas [34]. Protein molecules can be used to coat the NPs and protect them from the medium, to join multiple NPs in order to form higher order supramolecular assemblies, they can be introduced into hollow structures of the NPs or be surrounded by them [35]. The variety of conformational states of certain disordered and self-aggregating proteins may be exploited to tune the properties of the hybrid material to different purposes.

A number of groups have strived to elucidate the modes of binding of αS to simple NPs made by different core materials, including silica, gold and lipids, which exhibit biocompatibility and can be easily functionalized [36–38]. A variety of experimental and computational techniques were applied to gain information at the molecular and sub-molecular level on the organization of αS molecules within a hard corona, on the NP-induced structural transitions, the determinants of binding and the dynamic exchange processes at the nano-bio interface. The possibility to perturb the aggregation behavior of αS by use of NPs has attracted considerable interest [39,40]. Research efforts were largely focused on the observation of aggregation kinetics curves of αS in the presence or absence of NPs, providing insight into the aggregation process at the macroscopic level [41]. A mechanistic description of the effects of NPs on the aggregation kinetics has lagged behind, however few reports have provided preliminary insight into the microscopic events and structural conversions taking place during αS aggregation [42–44]. Interestingly, certain NPs were shown to interact with and disassemble preformed amyloid fibrils [40]. Tailored interactions between αS and NPs were further explored to develop reactive agents and nanobiocomposites featuring novel attributes for application in nanotherapeutics, nanooptics, nanoelectronics and other areas [45,46]. A large number of studies have been carried out to explore how NPs bind to and influence the aggregation propensity of diverse amyloidogenic proteins and peptides other than αS [22], however αS appears to have attracted greater attention than other IDPs for developing composite nanomaterials, presumably due to its unique favorable properties, such as molecular size, solubility, stability and ease of production.

The objective of this review is to summarize progress made in the study of αS interactions with NPs, with an emphasis on the conformational plasticity of the protein and its self-assembling propensity. The review is divided into three sections discussing: (1) fundamental aspects and molecular determinants of αS adsorption onto NP surfaces; (2) efforts aimed at controlling αS self-aggregation, formation of toxic assemblies and disaggregation of insoluble fibrils; (3) achievements towards the fabrication of αS-based hybrid materials presenting novel functionalities. We aim to provide the basis for better understanding the conformational properties of αS at the interface with NPs and illustrate how this knowledge may support our ability to control αS structural transitions and to design functional nanocomposites. Among the large variety of known NPs, we decided to focus our survey on the simplest type, namely particles of near-spherical shape. The selected case studies comprise both inorganic and organic materials as well as lipid nanovesicles. While the former types are attractive tools for exploratory purposes and applications, the latter are included for their great relevance as biomembrane mimics to probe αS conformational versatility and membrane surface-induced structural transitions.

2. Adsorption of Monomeric Alpha-Synuclein onto Nanoparticles

2.1. Silica Nanoparticles

The distinct physicochemical environment of the hard and the soft corona is expected to influence differently the conformational preferences of protein molecules. Recent work by Grandori and coworkers focused on the characterization of the conformations of αS and other proteins in the hard corona of silica NPs (SNPs) [37]. To prepare corona-coated SNPs, SNPs (~50 nm) were incubated with excess αS and then subjected to centrifugation and washing cycles. Based on transmission electron microscopy (TEM) analysis, the αS corona was found to be formed by a monolayer of tightly bound, collapsed molecules. Circular dichroism (CD) and Fourier-transform infrared (FTIR) spectra suggested the formation of helical segments on adsorption of αS to SNPs, however the effect was rather limited, indicating that the disordered state remained prevalent. Despite the limitations of the single experimental techniques in the analysis of protein-NP hybrids, such as possible scattering effects affecting CD experiments, the work demonstrated that a combination of them could provide useful structural insights into adsorbed protein layers.

To complement the knowledge acquired on the hard corona, a subsequent study was aimed at the characterization of αS molecules in dynamic exchange with the surface of SNPs [38]. Tira et al. used nuclear magnetic resonance (NMR) spectroscopy to gain insight into the adsorption mechanism at single-residue resolution. The direct observation of NMR signals from NP-bound proteins is generally unfeasible due to excessive line-broadening caused by the slow rotational tumbling of the hybrid species, however perturbations may be detected as intensity losses or exchange-averaged observables [47,48]. Indeed, heteronuclear single quantum correlation (HSQC) spectra and relaxation rate measurements performed on samples containing αS and SNPs revealed that the amphipathic amino-terminal domain was the primary contact with the NP surface, while the carboxy-terminal domain retained significant mobility (Figure 1A,B). An oxidized form of αS, containing four methionine sulfoxides, associated with oxidative stress, was found to exhibit similar binding properties as the unmodified species, indicating that the increased hydrophilicity did not influence the binding to SNPs. Interestingly, αS interacted with the surface of SNPs also in the molecularly crowded environment of blood serum with similar orientation as in simple buffer (Figure 1C). Additional competition binding experiments, supported by sodium dodecyl sulfate-polyacrylamide gel electrophoresis (SDS-PAGE) analysis, showed that αS was able to displace serum albumin and other proteins from the surface of NPs, while the highly basic four-repeat domain of Tau, an amyloidogenic IDP, displayed stronger affinity to SNPs, compared to αS. Thus, αS adsorption was described as a dynamic process wherein molecular exchange on the surface determines the composition and organization of the protein corona.

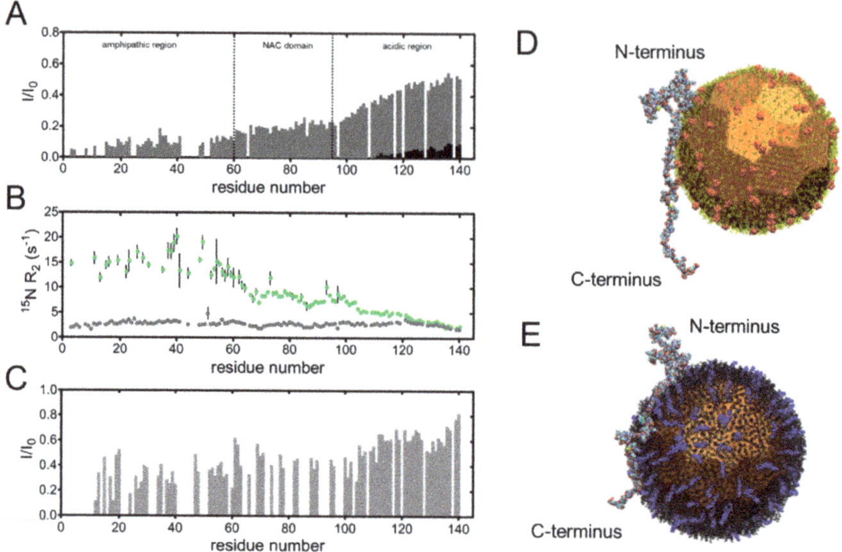

Figure 1. Orientation of αS molecules on the surface of nanoparticles. (**A–C**) Site-resolved nuclear magnetic resonance (NMR) interaction profiles revealing amino acid residues of αS involved in binding to silica nanoparticles (SNPs) on the basis of intensity attenuations or increased nuclear spin relaxation rates. (**A**) Relative signal intensities obtained from heteronuclear single quantum correlation (HSQC) spectra measured on ^{15}N-enriched αS dissolved in buffer solution, in the absence of SNPs (I_0) or in their presence (I), at two different concentrations (gray and black bars). (**B**) ^{15}N transverse relaxation rate values measured in the absence (gray dots) and presence of SNPs (green dots). (**C**) Relative HSQC signal intensities measured on ^{15}N-enriched αS in human blood serum containing (I) and not containing (I_0) SNPs. Adapted from ref. [38], Copyright 2020, with permission from Elsevier. (**D,E**) Rendering of simulated interaction of αS with AuNPs showing the reversed orientation of the protein depending on the capping ligand. (**D**) Interaction with citrate-capped AuNPs. (**E**) Interaction with MTAB-capped AuNPs. Citrate and MTAB ligands are charged (highlighted in red and blue, respectively). Adapted with permission from Lin et al. J. Phys. Chem. C Nanomater. Interfaces. 2015, 119(36), 21035–21043. Copyright 2015 American Chemical Society.

The capability of nuclear magnetic resonance (NMR) spectroscopy to provide atomic-resolution information on protein-NP interactions and the qualitative observation that the binding of IDPs to certain NPs is related to the polypeptide sequence, stimulated Brüschweiler and coworkers to attempt a quantitative analysis of residue-specific NMR data [49]. The authors developed an interaction model based on a quantitative NP affinity scale determined from measurements on single amino acid types. In conditions of rapid exchange between free and surface-bound states, the difference in residue-specific ^{15}N-spin transverse relaxation rate constants, ^{15}N-R_2, observed in the presence and absence of NPs, termed ΔR_2, provides a direct measure of the interaction strength between the NP and residues in the polypeptide chain. After a number of refinements, the authors came up with a binding model capable to accurately predict residue-specific binding affinities of αS and other IDPs to SNPs [50,51]. This work provided mechanistic insight into the binding of αS and SNPs, explaining the observed interaction profile in terms of the non-uniform distribution of charged and neutral amino acids as well as in terms of global and local cooperativity effects.

Overall, the binding of αS to SNPs appears as a simple reversible two-state binding mechanism mediated in large part by the lysine-rich N-terminal domain, which experiences attractive electrostatic interactions with the negatively charged, deprotonated silanol groups at the SNP surface. However, this simplified picture does not explain all observations, such as the apparently distinct involvement of the C-terminal domain at different protein/NP ratios and the formation of a slowly desorbing hard corona [38]. Furthermore, the role of intermolecular interactions in the structural organization of the protein corona have remained elusive. Diverse techniques have served to provide detailed descriptions of the organization of folded protein molecules bound to NPs [52–54], however similar results were not obtained for αS, likely due to the greater difficulty of obtaining clear data for an unstructured polypeptide. All of these aspects deserve attention in future studies.

2.2. Gold Nanoparticles

The study of the adsorption of proteins to AuNPs has invariably involved the use of capped or functionalized AuNPs. In an early work, Murphy and coworkers investigated the interaction of αS with citrate-capped 20 nm and 90 nm AuNPs [55]. The authors used dynamic light scattering (DLS) to monitor changes in the mean hydrodynamic diameter after mixing the protein with the NPs. They observed the formation of a relatively thick, strongly bound adlayer (hard corona) and a less thick, labile soft corona. An overall apparent binding constant of $(2.0 \pm 0.4) \times 10^7$ M^{-1} was estimated from the DLS data using 20 nm AuNPs and a similar result was obtained from the analysis of plasmon band maxima in the UV–vis spectra. Fluorescence quenching was exploited to separately quantify the binding constants for the hard and the soft corona, yielding affinity constant values in the order of 10^7 M^{-1} and 10^{-3} M^{-1}, respectively. The latter value suggested that the binding of the soft corona was thermodynamically unfavorable and kinetically driven. The authors further attempted a structural analysis of NP-bound protein by CD using a stacked double-cuvette method but realized that high absorbance by the metallic core could compromise the quality of the obtained results. Therefore, they resorted to use enzymatic digestion followed by mass spectrometry (MS) analysis to gain insight into the structure of αS in the hard corona. A comparison of trypsin digestion patterns of free αS with bound αS on 20 nm citrate-capped AuNPs suggested that the protein maintained its native unstructured state when bound on AuNPs, with the N-terminal section strongly adsorbed onto the NP surface.

As opposed to citrate-capped AuNPs, which expose a negatively charged surface, poly (allylamine hydrochloride) (PAH) coated AuNPs display a positively charged surface. Murphy's group used a similar methodology as that used with citrate AuNP to explore the structure of αS bound to PAH AuNPs [36]. The protein was found to adsorb as multilayers when present at low protein/NP ratios and eventually formed agglomerates at higher ratios. The latter condition was attendant with an increase in β-sheet structure and decrease in α-helical content, possibly explaining the tendency to form agglomerates. Apparently, the mode of adsorption could elicit the seeding of a global conformational change of αS in the sample. Based on trypsin digestion data, the protein molecules were found to adopt random orientation in the multilayered corona.

Solution NMR spectroscopy was used in a subsequent study to obtain definitive insight into the orientation of αS on both anionic and cationic AuNPs [56]. For cationic particles, in spite of using PAH, AuNPs were capped with (16-mercaptohexadecyl) trimethylammonium bromide (MTAB), a ligand that did not promote protein aggregation and therefore allowed better interpretation of the NMR data. As expected, portions of the protein that bound to NPs exhibited larger linewidths and attenuated signals. A comparison of the residue-by-residue intensity profiles collected with the anionic and cationic NPs clearly showed the reverse orientation of the protein, with the prevalently basic N-terminus acting as the anchor to citrate-capped AuNPs and the acidic C-terminus being prevalently bound to MTAB AuNPs. Besides identifying the sections in direct contact with the NPs' surfaces, the NMR data also revealed that the unanchored portions experienced restricted motion due to their tethered condition. In both cases, the protein remained disordered upon binding to the NPs. Molecular dynamics (MD) simulations supported the observed reversal of protein binding orientation (Figure 1D,E) and additionally indicated that the central hydrophobic segment, the NAC domain, was attracted to both types of AuNPs. These results demonstrated the possibility to obtain molecular control of protein display on engineered NPs.

2.3. Lipid Nanovesicles

In cells, αS partitions between disordered or partly ordered cytosolic forms and phospholipid-bound states [57–59]. The association of αS with phospholipid membranes was linked to a role of the protein in the regulation of reserve pools of synaptic vesicles and dopamine homeostasis [6,60]. Such crucial function has spurred the investigation of αS binding to lipid nanovesicles as membrane mimics. The N-terminal 11-mer repeat sequence of αS, resembling the sequence motifs found in apolipoprotein A-I, suggested that it could form amphipathic helical lipid-binding domains [61]. Indeed, several in vitro studies have established that αS undergoes a coil-to-helix transition when binding to vesicles made by acidic phospholipids or anionic detergents [61–63].

In studies of αS-lipid interactions, a large variety of lipid/detergent vesicles of different composition and size have been used and several distinct experimental conditions were tested, which may in part account for the fact that results were sometimes seemingly contradictory [2,64]. The increase in helical content on binding of αS to anionic phospholipid vesicles was initially observed by CD spectroscopy and the interaction was proposed to be mediated by four N-terminal helices (region 1–60) [61]. In a subsequent study, Eliezer and coworkers used small (~5 nm) SDS detergent micelles as membrane mimics in place of larger phospholipid vesicles to facilitate observation of the lipid-bound state by NMR spectroscopy [12]. They found that binding of α-synuclein to SDS micelles elicited formation of an extended α-helix encompassing residues 1–100, while the C-terminal portion of the protein remained unassociated. Later investigations, based on NMR and partial tryptic digestion, identified a short break within the extended helical region [63,65]. Furthermore, spin probe-induced broadening of NMR signals, ^{15}N relaxation measurements and fluorescence spectroscopy data indicated the presence of two N-terminal helices, positioned on the surface of the SDS micelle and separated by a flexible stretch [66]. The region of residues 61–95 was found to adopt a helical conformation but it was proposed to be partially embedded in the micelle [66]. Finally, a high resolution structural determination established that micelle-bound αS forms two curved α-helices within the N-terminal domain, connected by an ordered, extended linker in an anti-parallel arrangement, followed by another short extended region and a largely disordered tail (residues 98–140) [67].

SDS micelles have provided an invaluable system to study structural properties of lipid-bound αS, however both the chemical composition and the size do not entirely recapitulate the features of the ~50 nm presynaptic phospholipid vesicles. Thus, several studies have been conducted using either small or large unilamellar vesicles (SUVs or LUVs). A structural model of SUV-bound αS was obtained by applying simulated annealing MD restrained by the immersion depths and long-range distances obtained from continuous-wave and pulsed electron paramagnetic resonance (EPR) data [68] (Figure 2). The bound form was described as an extended helix (ca. 90 amino acids long) with a curved arrangement that follows the curvature of the vesicle surface and allows lysine residues to interact with phosphate groups, acidic residues to approach the choline groups and hydrophobic residues to associate with the lipidic moieties. Further experimental evidence, based on single molecule Förster resonance energy transfer and on the use of 100 nm LUVs, supported the extended helix model [69], however other authors concluded that the broken-helix arrangement best described the SUV-bound protein state [70]. Later studies provided evidence for coexisting populations of broken and extended helices, in part reconciling the divergent models [71,72]. It was found that relative protein/lipid concentrations and vesicle size could modulate the preference of αS for distinct helical arrangements. A multiplicity of coexisting binding modes were further proposed by Bax and coworkers, featuring slow exchange kinetics and involvement of N-terminal segments of differing length [73].

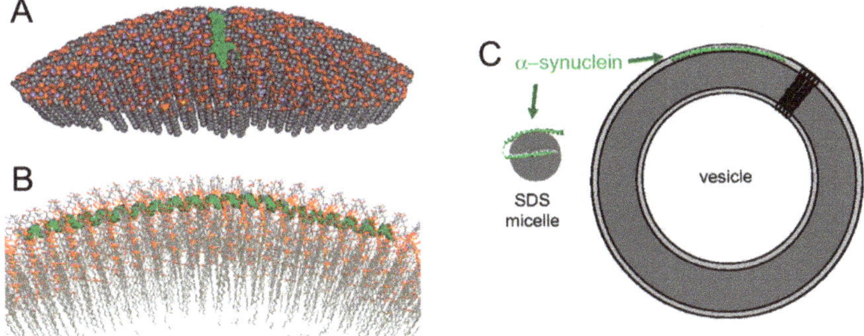

Figure 2. Interaction of αS with curved lipid surfaces. (**A**) Space-filled model of α-synuclein (shown in green) binding to the surface of a lipid vesicle 300 A in diameter; ≈ 25% of the outer leaflet of the vesicle is shown. The vesicle was fitted around one of the structures derived from the experimentally restrained simulated annealing—molecular dynamics (SAMD) calculations. (**B**) A closer cross-sectional view of the αS interaction with the lipid surface, with rotation through 90° from the image in A. The protein (green) follows the curved surface of the vesicle, with the helical axis positioned just below the level of the phosphate groups of the lipids. This position of the protein emerged from the SAMD calculations and reflects the immersion depths obtained from the continuous-wave electron paramagnetic resonance (EPR) data. (**C**) Cartoon representations of the structures of α-synuclein on micelles and small unilamellar vesicles (SUVs). The small and highly curved micelles cannot accommodate the extended helical structure present on the membrane. Adapted with permission from Jao et al., Proc. Natl. Acad. Sci. USA, 2008, 105 (50) 19666–19671. Copyright 2008 National Academy of Sciences.

Despite the complexity of lipid vesicle binding, consensus was reached about the preferential interaction of αS with anionic phospholipids, although non-electrostatic interactions with neutral and zwitterionic lipids were also observed. A systematic study performed with phospholipid bilayer nanodiscs, an alternative model of lipid membranes exhibiting planar surfaces, confirmed the electrostatic model and showed that the binding mode was dependent on the relative abundance of anionic lipids versus neutral molecules [74]. Additionally, the degree of saturation and the length of the acyl chains were found to influence the binding of αS [75,76]. More specifically, molecular properties that determine the lipid phase state and membrane fluidity critically influence αS adsorption [74]. Interestingly, αS and anionic phospholipids may also form nanometer-sized lipoprotein particles, reminiscent of high-density lipoproteins, in which αS adopts a helical secondary structure [77]. Thus, the specific lipid environment has a profound impact on the partitioning and conformational transitions of αS, suggesting the possibility to tune molecular properties and regulate biomolecular and nano-bio interactions by careful design of the lipid-based nanomaterials.

2.4. Mixed-Type and Other Nanoparticles

The widely documented attraction of synuclein to lipid layers has inspired the development of lipid-based composite particles, other than simple vesicles. For example, Lee and coworkers used osmotic shock to coat monodisperse SNPs (60 nm diameter) with a lipid membrane, thereby obtaining spherical NP-supported lipid bilayers (SSLBs) [78]. Specifically, SNPs were amine-functionalized and the lipid coating was made by a mixture of the anionic lipid DOPA (1,2-dioleoyl-sn-glycero-3-phosphate) and the zwitterionic lipid DOPC (1,2-dioleoyl-sn-glycero-3-phosphocholine). SSLBs offer important advantages over SUVs/LUVs as they display larger X-ray scattering cross section of the silica core relative to membranes, enabling small-angle X-ray scattering (SAXS) and state-of-the-art X-ray photon correlation spectroscopy (XPCS), thereby expanding the repertoire of experimental techniques available to probe colloidal structure and dynamics of αS-bound vesicles. The authors reported that αS disrupted vesicle-vesicle interactions, with implications for synaptic membrane fusion and the ultrastructure and dynamics of synaptic vesicle pools.

The coating of inorganic NPs with lipid layers was also recently pursued by other groups [79]. The surface of AuNPs of varying size was capped with an inner layer of dodecanethiol and an outer layer of SDS. Thus, the organic bilayer was formed on a rigid scaffold and was resistant to deformation by αS, as opposed to lipid vesicles which are known to undergo structural remodeling upon interaction with the protein. By removing the effect of NP deformation, the obtained mimics allowed investigating the effects of NP curvature on protein binding behavior. SDS-AuNP—bound αS displayed increased solvent accessibility in the NAC region, suggesting that the adsorbed protein could possess higher aggregation propensity than unbound αS.

In the study of potential applications of NPs as protein aggregation modulators, Hajipour and coworkers focused on the interaction between αS and graphene sheets and superparamagnetic iron oxide NPs (SPIONs) with different surface properties and sizes [44]. Graphene sheets were prepared as small (150–250 nm), medium (450–650 nm) and large (800–1200 nm) sheets with polyglycerolsulfate, polyglycerol and polyglycerolamine coverages, displaying surface charges of −30 mV, 0 and +30 mV, respectively. The investigated SPIONs were 20, 50 and 100 nm large and displayed various functionalities (COOH, NH_2, PEG-300 and chitosan). Atomistic MD simulations indicated that the interaction of αS with charged nano-objects was initially driven by electrostatic attraction, with a later involvement of hydrophobic residues and non-polar contacts dominating the interaction. Additionally, hydrogen bonds were formed, particularly in the case of amine-functionalized graphene. Differences in αS binding behavior to graphene and SPIONs were attributed to the distinct shape and corresponding surface curvature of the two materials. The affinity to neutral nano-objects was lower compared to charged ones. The various binding contributions resulted in distinct orientations of αS on the surface of different particles, supporting the view that distinct nanomaterials could differently affect αS self-assembly at the nano-bio interface.

Given the application of human serum albumin (HSA) NPs as carriers for drug delivery into the brain, HSA NPs were studied by Otzen and coworkers in order to assess the nature of their interaction with αS [80]. HSA NPs were produced as unmodified or polyethyleneimine(PEI)-functionalized ~35–40 nm spherical particles. By use of a centrifugation assay, fluorescence anisotropy measurements and small-angle X-ray scattering (SAXS) data, the authors found that αS was attracted much more strongly to positively charged PEI-HSA NPs than to negatively charged HSA NPs. Changes in the CD spectral shape suggested that the interaction with PEI-HSA NPs caused a conformational change in αS. The absence of a significant binding to HSA NPs, despite their net negative surface charge, may suggest that interactions with biological NPs are distinct relative to most synthetic platforms because the former exhibit an inhomogeneous distribution of polar, charged and nonpolar groups as opposed to the generally isotropic presence of chemical groups on the surface of common synthetized NPs.

3. Influence of Nanoparticles on Alpha-Synuclein Aggregate Formation

3.1. General Surface Effects on Aggregation Kinetics

αS fibrillation is commonly considered a nucleation-dependent growth process that follows sigmoidal kinetics with distinct phases of nucleation, propagation and equilibration [81,82] (Figure 3A,B). The primary nucleation step (commonly associated with the lag-phase), during which soluble species are sequestered into oligomers of different sizes and structures, is much slower compared to the addition of monomers to preformed protofibrils, which leads to an exponential growth of fibrils (growth, propagation or elongation phase) until a state of equilibrium is achieved (plateau phase). This simplified two-step model adequately describes the evolution of the system from a macroscopic point of view, however it does not consider the multitude of microscopic events that contribute to the entire process. It is now established that simultaneous microscopic processes are ongoing during all phases and that processes other than primary nucleation and growth, such as fibril fragmentation or secondary nucleation, may represent important events during aggregation [83–85]. On-pathway and off-pathway intermediate supramolecular assemblies exhibit different degrees of cytotoxicity, however their identification and characterization remains challenging [81,86]. Due to the complexity of the aggregation phenomenon and the difficulty to access microscopic events, most studies involving perturbations of αS aggregation focus on comparative analyses of the kinetics at the macroscopic level. Experimentally, kinetics curves are generated by following time-dependent changes in sample turbidity or the fluorescence intensity of fibril-responsive dyes, such as Thioflavin T or by means of several other biochemical and biophysical techniques [21].

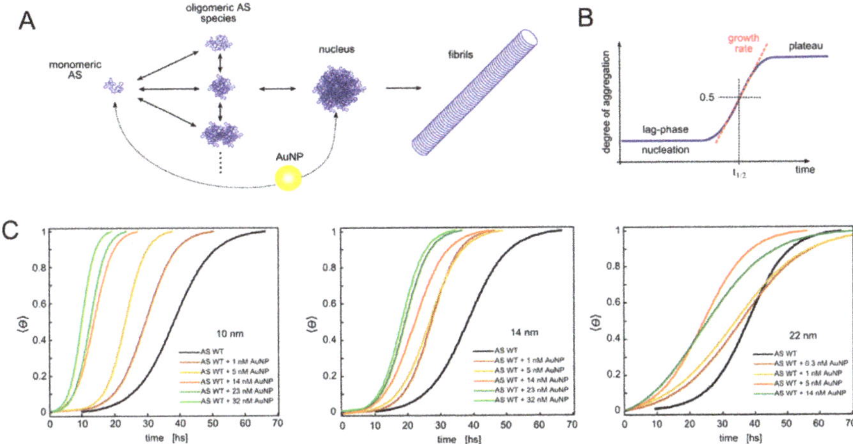

Figure 3. Nanoparticle-induced perturbation of fibril formation kinetics. (**A**) Schematic of the aggregation process of αS. AuNPs may interact with monomeric and/or oligomeric αS during the process of nucleation. (**B**) Diagram of the sigmoidal kinetics corresponding to nucleation followed by fast, autocatalytic growth. (**C**) Average θ(t) for samples with different concentrations of AuNPs with diameters of 10, 14 and 22 nm, respectively. θ(t) is the aggregate-responsive spectroscopic observable. Adapted with permission from Álvarez et al. Nano Lett, 2013, 13 (12) 6156–6163. Copyright 2013 American Chemical Society.

Surfaces have a profound impact on aggregation phenomena [42,87]. It has long been recognized that lipid membrane surfaces can act as catalysts of fibril formation, serving as a platform for nucleation and further polymerization [88,89]. Actually, membrane binding of αS appears to be an important factor in the pathogenesis of Parkinson's disease [90]. Membrane-assisted aggregation could result from several concurrent factors, including a reduction of conformational entropy in the bound state, the induction of structural ordering, surface molecular crowding effects and lowering of the local dielectric constant which may facilitate the formation of intermolecular contacts [91–93]. The notion that surfaces may facilitate fibril formation is frequently exploited in in vitro aggregation studies of certain IDPs, including αS, which display extremely slow aggregation kinetics in solution. The addition of beads or other surfaces imparts a dramatic acceleration on protein self-assembly, allowing to perform experiments in practical time frames. Fibril seeds themselves offer particularly active surfaces that catalyze nucleation of aggregation-resistant proteins. Based on these evidences, NPs have attracted much interest owing to their high surface-to-volume ratio [94], compared to bulk material and many studies have been carried out on the effects of NPs on the aggregation of amyloidogenic proteins [22]. It emerges that NPs can lead to either acceleration or retardation of the fibril formation, depending on several factors such as the physicochemical properties of the NPs (e.g., size, charge, nature of exposed chemical groups), the amino acid composition and stability of the IDP, the concentrations of the solutes and the characteristics of the solution (e.g., ionic strength, pH) [95]. Thus, proper tuning of the NP properties shows promise as a means to control fibril formation pathways and could eventually provide a novel strategy for therapeutic intervention against aberrant protein aggregation in neurological disorders. Nonetheless, it must be considered that NP properties will also determine their ability to cross the blood-brain barrier and the cellular membranes before reaching the neuronal cytosol. Available pathways and strategies adopted to deliver NPs across the blood-brain barrier have been recently reviewed [96] and significant progress has been made in understanding the factors that control cellular uptake of NPs [97,98]. Most of the case studies presented here aimed at probing specific surface effects on protein aggregation and often did not address the ability of NPs to cross biological barriers.

A growing number of NP materials have been tested for their capability to influence the fibril formation process of αS. They include metals (Au, Fe) [41,99,100], metal/metalloid oxides (CeO_2, TiO_2, ZnO, Fe_3O_4, SiO_2) [37,38,44,101–104], carbon (graphene, fullerenol) [44,105,106], polymers (dendrimers, others) [42,107–111], biomolecules (protein-based) [80,112] and lipids (detergent, phospholipid) [74, 113–115]. The diversity of sizes, surface groups, charge density, hydrophobicity/hydrophilicity and other features of the investigated particles, render the rationalization of the results a challenging task. For example, both negatively and positively charged NPs were reported to inhibit αS fibrillation and both types were found to accelerate the process. In this regard, it appears that in order to establish more general paradigms, studies on a single protein/NP pair are of limited use, while more systematic investigations would be more informative. Furthermore, we emphasize the importance of conducting kinetics experiments with extreme rigor (e.g., avoiding NP precipitation, performing several replicas) and of carefully reporting the conditions in which experiments were performed (solution conditions, solute concentrations, mechanical agitation, etc.), to ensure that results can be compared and interpreted correctly.

A systematic study of the interaction between αS and citrate-capped AuNPs in a range of sizes (10, 14 and 22 nm) and of concentrations was carried out by Stefani and coworkers [41]. To monitor the evolution of the aggregation process at a macroscopic level, the team adopted an approach based on the use of the environment-sensitive fluorescent probe MFC, covalently attached to an engineered αS. This dual-emission dye is a multiparametric fluorescent probe, highly sensitive to changes in polarity and hence an exquisite reporter of early protein aggregation events [116]. The aggregation kinetics of αS in the presence of NPs followed a sigmoidal trend, exhibiting a transition halftime that decreased with the increase of AuNP concentration (Figure 3C). The overall acceleration produced by the AuNPs could be traced back to distinct effects on the nucleation and growth phases. For smaller AuNPs, the growth rate increased with particle concentration, while such effect was not observed with the 22 nm AuNPs. Instead, the latter had a more pronounced effect on the duration of the lag phase. The observed effects on the transition halftime and the growth rate did not scale with the available surface area. The authors attributed the reduction of the lag phase to the accumulation of protein on the AuNP surface, promoting the formation of critical nuclei for fibrillization. They further explained the variations in growth rate as a function of AuNP size by the formation of nuclei of different nature. This study therefore illustrates that the impact of NPs on the aggregation of αS is often not linear with some property of the NPs but multifaceted, underscoring the complexity of microscopic interactions occurring at the nano-bio interface.

3.2. Mechanistic Insights into Nanoparticle-Mediated Perturbations of αS Aggregation

A mechanistic description of the effects of surfaces on the aggregation kinetics of polypeptides requires consideration of the concurring microscopic events. In this respect, several available strategies may prove useful, such as the global analysis of macroscopic aggregation curves measured under different conditions [117], computational methods that take into account the structural plasticity of the polypeptide as well as the nature of the NP surface [117,118] and experimental procedures that allow the observation of structural transitions or the detection of transient intermediates. To date, a limited number of such studies have been applied to αS/NP systems.

The consideration that certain NPs accelerate protein aggregation, while others cause a retardation of fibril formation, prompted Linse and coworkers to explore which factors were responsible for one or the other behavior [42]. By use of a dynamic Monte Carlo method, the group simulated amyloid growth profiles in the presence of surfaces with varying attraction potential. The results showed that weakly attractive surfaces (peptide binding constant, $K = 0.0017\ \mu M^{-1}$) determined a reduction of the nucleation rate, compared to non-attractive surfaces, while highly attractive surfaces ($K = 0.16\ \mu M^{-1}$) accelerated the nucleation process. Indeed, weakly attractive objects simply reduce the concentration of free monomers available to form aggregation nuclei in solution but they do not change the microscopic events. By contrast, on the surface of highly attractive materials, distinct nuclei are formed in addition to those that form in solution. In the case of an intermediate attraction potential, the apparent kinetics was similar to that occurring with non-attractive surfaces, however the structures of the initial aggregates were different. Interestingly, the above observations were modulated by the intrinsic properties of the polypeptide, whereby for example weakly attractive surfaces retarded fibril formation by aggregation-prone mutants but accelerated the process for more stable polypeptides. ThT fluorescence assays performed using plain polystyrene NPs and αS consistently showed an acceleration of fibril growth upon increasing NP concentration, as a consequence of a strong non-electrostatic attraction causing surface-catalyzed nucleation. In general, it seems possible that positively and negatively charged NPs could exert the same inhibitory or acceleratory effect on the fibrillization of αS, because they would establish electrostatic interactions with either the acidic C-terminal or the basic N-terminal domains [56]. Neutral objects were reported to have limited impact on the aggregation rate, presumably because they were not able to interact significantly with αS [44].

While computational methods can capture the formation of early-stage aggregates and provide atomistic insights into conformational transitions with relative ease, the identification of transient intermediates and the detection of time-dependent structural changes during fibrillization remain experimentally challenging. Chattopadhyay and coworkers applied fluorescence correlation spectroscopy (FCS) and laser scanning microscopy (LSM) to investigate the early events of αS aggregation in the presence of pristine or Lys-modified Fe_3O_4 NPs [43]. The use of these techniques proved advantageous over the use of standard dye-based methods which are quite insensitive to the formation of smaller aggregation intermediates. Bare Fe_3O_4 NPs were found to accelerate early and late-stage aggregation of αS, while Lys-coated Fe_3O_4 NPs displayed an inhibitory effect. Maximum entropy analysis of the correlation functions measured by FCS detected increases in conformational heterogeneity during the progress of aggregation and revealed the presence of aggregated species at earlier time points in the presence of bare Fe_3O_4 NPs. The early aggregates were visualized by LSM. The molecular basis for the different perturbations elicited by bare and Lys-coated particles remains elusive, however the study demonstrated the possibility of acquiring important information on aggregates in heterogeneous systems.

Hajipour and colleagues used size exclusion chromatography coupled with multi-angle light scattering (SEC-MALS) to quantify the amount of small αS oligomers formed in the presence of SPIONs or graphene nanoobjects and ranked the particles according to their ability to trigger the formation of such aggregates [44]. They were further able to determine whether the formed oligomers were on- or off-pathway with respect to fibrillization. In another study, Tira et al. followed the time course of αS aggregation in the presence of SNPs using CD spectroscopy [38]. The authors could track the time-dependent structural transitions of the polypeptide from its native disordered state to conformations with mixed α and β secondary elements. Interestingly, the shapes of CD spectra collected at different time points for αS and its methionine-oxidized form in the presence of SNPs indicated that the two species formed distinct supramolecular assemblies, consistent with the reported resistance of the oxidized species to form fibrils [119]. Taken together, the information obtained from experiments on prefibrillar species could be used to inform computational methods and contribute to an improved description of the complex aggregation pathways.

3.3. Lipid Surface-Mediated αS Aggregation

The nature of lipid surfaces is quite unique due to their soft, dynamic character which allows peptides to penetrate into the lipid layer(s), at least partially [120,121]. In addition, lipids and amyloidogenic polypeptides share an amphipathic structure and are therefore intrinsically prone to interact. The interactions are modulated by several factors and are highly dependent on lipid composition, surface charges and thermotropic properties [122,123]. Hence, the mechanisms by which lipid nanovesicles affect αS fibril formation may depart from those involved with other nanomaterials. Indeed, the mode of association with lipid surfaces modulates αS aggregation in different ways [91,115,124,125]. Furthermore, it has been suggested that fibrillization in the presence of lipid molecules may result in the formation of protein-lipid co-aggregates [126–128] and that the morphology of fibrils is modulated by the relative proportion of protein and lipids [129].

αS partitions dynamically to SUVs and LUVs composed of anionic phospholipids such as phosphatidylserines [39,123]. Interestingly, the binding affinity is highest when the lipid layers are in the fluid state, as opposed to the gel-like state, since in the former case the hydrophobic portions of the lipid molecules are on average more exposed [123]. In turn, αS binding affects the lipid phase behavior and induces lipid segregation into protein-poor and protein-rich populations [123]. Yet, unexpectedly, the lipid phase state does not correlate with the vesicle-promoted acceleration of αS amyloid fibril formation [123]. Instead, the kinetics of aggregation was found to correlate with the solubility of the lipid molecules, suggesting that at least part of the free energy barrier for the aggregation process is associated with the translocation of lipid molecules from a membrane to a protein environment [123].

Aggregation assays carried out under quiescent conditions and at varying dimyristoyl phosphatidylserine (DMPS)/αS ratios indicated that αS did not convert into fibrils when excess lipid was present and most of the protein was in the bound state [39]. On the contrary, at lower lipid concentrations, when significant populations of both free and lipid-bound protein were established, SUVs determined the rapid formation of amyloid fibrils. Indeed, a combined experimental and theoretical analysis indicated that at low lipid/αS ratios, the bound protein promoted a primary nucleation process, much faster than that occurring in bulk solution. The facilitated nucleation was attributed to the high local concentration of protein on the SUV surface and to a conformational shift towards aggregation-competent states [39]. Under the used conditions, other microscopic processes, including homogeneous primary nucleation, secondary nucleation and fragmentation, did not contribute measurably to the aggregation reaction. A similar finding was reported concerning the effect of nanovesicles made by zwitterionic lipids, which influenced the lag time more than the fibril elongation rate [113]. Under quiescent conditions, secondary processes may be prevented due to kinetic trapping of the lipid-bound fibrils [39]. Site-resolved NMR data obtained using nanodiscs as membrane models indicated that region-specific membrane affinities (particularly of the NAC region) were correlated with aggregation behavior [74]. The dual effect of lipid surfaces to both accelerate or inhibit αS amyloid fibril formation depending on the relative proportion of protein and phospholipids may suggest a possible mechanism for the onset of aberrant aggregation as a consequence of altered levels of αS expression, associated with some forms of PD [130].

It has been proposed that the transient interaction of αS with lipid bilayers may determine the formation of a pool of helical conformers that are aggregation-resistant [131]. Thus, physiologically, lipid surfaces may act as chaperones that assist the folding process of otherwise disordered protein molecules. An imbalance in the relative populations of protein in the folded and unfolded pools may cause aberrant aggregation in pathology. These findings suggest the possibility to develop tailored NPs as artificial cofactors that assist the formation of aggregation-resistant αS species.

3.4. Nanoparticle-Fibril Interactions and Fibril Disassembly

The removal of amyloid deposits is a prominent therapeutic aim in protein misfolding diseases [132]. However, the disassembly of preformed amyloid fibrils is both challenging and risky. On the one hand, protein fibrils are insoluble, extremely stable and resistant to degradation. On the other hand, disaggregation may exacerbate amyloid toxicity by increasing the load of toxic oligomers [133]. Thus, in order for NPs to be effective in reverting aberrant deposition of protein fibrils, they must display significant affinity for the fibrillar structures, establish interactions that weaken or disrupt the dense network of hydrogen bonds that stabilize the stacked β-strands and promote the conversion of fragments into harmless products. Diverse NPs were shown to interact with preformed αS fibrils.

Citrate-capped AuNPs of the size of 22 nm but not of 14 nm and lower, were found to associate with fibrils, indicating a size-dependent affinity [41]. Mercapto-undecanesulfonate-coated AuNPs were developed to target synthetic, recombinant and native fibrils derived from different amyloidogenic proteins, including αS [134]. Such particles did not exhibit fibril-disaggregating properties, instead they were exploited to label amyloid fibrils for assessing morphological polymorphism using cryogenic electron microscopy (cryo-EM) [134]. Both SPIONs and graphene were reported to disassemble αS fibrils, with a higher efficacy shown by positively charged nano-objects, possibly related to their higher affinity to charged αS residues [44]. Upon fibril fragmentation, the amount of oligomeric species did not increase, suggesting a safe use of these NPs [44].

In addition to graphene, also graphene quantum dots (GQDs) were shown to induce dissociation of αS fibrils [40]. Ko and coworkers carried out a thorough characterization of this process [40]. GQDs produced the dissociation of fibrils into short fragments of an average length of 235 nm and 70 nm after 6 and 24 h, respectively. After 3 days of incubation, the number of fragments decreased, indicating that the process progressed until complete disassembly. The interaction of the negatively charged GQDs with αS was likely initiated by electrostatic attraction with the protein's N-terminal domain. MD simulations performed with the sole NAC domain indicated that after initial binding, the β-sheet structure of the outer monomer was rapidly and completely destroyed as a consequence of strong hydrophobic interactions between GQDs and valine residues. Importantly, GQDs could penetrate the blood-brain barrier and protect mice against dopamine neuron loss induced by preformed amyloid deposits.

Dendrimers have been recognized as potential powerful agents for the disaggregation of fibrils, displaying strong binding affinity to fibrillar structures and exerting their destructive effect by acting as efficient chaotropes [135]. Specifically, polyamidoamine (PAMAM) dendrimers were shown to interact with αS fibrils, the binding affinity increasing with the generation number (G4 < G5 < G6) [109]. The combined evidence from TEM, CD and ThT fluorescence indicated that dendrimers attacked the fibrils along the entire filament, not just at the ends and lead to amorphous aggregation. Similar to PAMAM dendrimers, also urea(U)- and methylthiourea(MTU)-modified polypropyleneimine (PPI) dendrimers were found to disaggregate preformed fibrils [136]. TEM detected the fragmentation of fibrils by G3-MTU-PPI into smaller and less organized aggregates. Interestingly, both types of NPs were found to significantly reduce the αS fibril load inside SK-MEL-5 cells in a dose-dependent and generation-dependent (MTU-PPI) or generation-independent (U-PPI) manner. However, MTU-PPI dendrimers displayed higher cytotoxicity in the presence of preformed αS fibrils [136]. Further studies on the effects of dendrimers and other polymeric particles on αS aggregation have been reviewed elsewhere [137].

4. Alpha-Synuclein/Nanoparticle Conjugates and Hybrid Nanomaterials

Functionalization or the combination of NPs with αS molecules has been intensely explored to produce nanobioconjugates and hybrid nanomaterials for diverse technological applications in bioanalytical chemistry and bionanotechnology [45,138,139]. Depending on the NP material and the aim of the application, a variety of conjugation strategies have been exploited, ranging from the simple deposition of the biomolecule on the NP surface [46,139], to non-covalent high affinity binding [45] and to covalent bond formation [138,140]. The conformational versatility of αS made it possible to exploit different properties of the associated molecules, from a disordered and highly dynamic form to the ordered superstructure typical of fibrils. Furthermore, protein self-assembly was exploited to fabricate ordered multi-component, supramolecular nanomaterials [46].

Jares-Erijman and coworkers explored the possibility to develop novel reactive agents or nanoactuators, based on the decoration of QDs with multiple copies of αS [45]. αS containing an A90C single point mutation was conjugated to biotin via maleimide chemistry, while QDs were capped with streptavidin. Thus, αS-QD nanoconjugates were obtained exploiting the formation of the well-known high affinity biotin-streptavidin complex. The nanoactuators were found to accelerate the formation of αS amyloid fibrils, both in vitro and in live cells, thereby acting as artificial nucleation seeds. The catalytic effect was attributed to the self-assembly of the protein initiated by the high local concentrations displayed at the surface of the NPs. The developed system could facilitate cellular studies of amyloid formation.

The production of AuNP-biomolecule conjugates often exploits the spontaneous formation of ligand monolayers via unique bonds between the Au surface and sulfhydryl groups of the biomolecule. Paik and coworkers found that a thin shell of modified αS molecules conjugated to AuNPs via Au-S bonds facilitated their deposition into a regular two-dimensional array on a glass support [138]. The obtained material constituted the basis for the production of a surface-enhanced Raman scattering biosensor. The platform was indeed responsive to phthalocyanine tetrasulfonate, an αS ligand and aggregation inhibitor and to metal ions forming complexes with the compound. The formation of a uniform array of coated AuNPs was attributed to the capability of fixed αS molecules to establish intermolecular interactions.

The αS-mediated assembly of AuNPs into hierarchical superstructures was further exploited to fabricate a flexible, free-standing NP monolayer film useful for the development of bio-integrated nano-devices and high-performance sensors [46] (Figure 4). Cysteine-free protein was first adsorbed onto AuNPs (10–30 nm) and subsequently onto a polycarbonate substrate. Fibril-like protein-protein linkages were revealed by scanning electron microscopy and β-sheet structure formation was observed by attenuated total reflectance Fourier transform infrared spectroscopy. The conversion into β structures was probably triggered by the exposure of the protein-AuNP system to chloroform, necessary to release the film from the support. Importantly, the process was specific for αS, as the film could not be obtained with other model protein molecules. An analogous concept was used to develop metal NP-based organic field-effect transistors with electrical memory [139]. A closely packed AuNP monolayer was obtained by the pH-dependent adsorption of αS-AuNP conjugates on a SiO_2 surface. Pentacene, a high-performance organic semiconductor, was finally deposited on the formed αS-AuNP film. The use of αS proved invaluable for the optimal controllability over the hybrid material structures, allowing to obtain highly tunable memory performance.

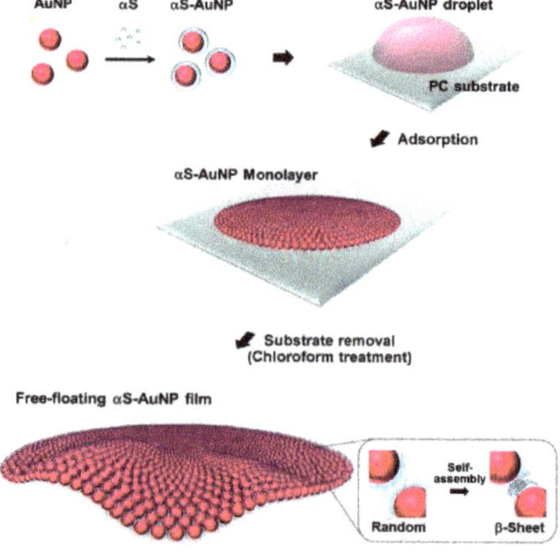

Figure 4. Production of functional αS-based nanobiocomposites. The fabrication of a free-floating

αS-AuNP monolayer film through the preparation of αS-coated AuNPs, adsorption of the αS-AuNPs onto the polycarbonate (PC) substrate and subsequent free-standing film production by αS–αS self-assembly upon removal of the substrate with chloroform. Reproduced with permission from Lee et al. Angew Chem. Int. Ed., 2015, 54 (15), 4571–4576, © 2020 WILEY-VCH Verlag GmbH & Co. KGaA, Weinheim.

Another study demonstrated the possibility to produce αS-based nanocomposites as intracellular drug delivery systems [140]. Mesoporous SNPs (~100 nm size and pore diameter of 2–3 nm) were coated with AuNPs, previously functionalized with αS, to yield 'raspberry-type' particles-on-a-particle (PoP) structures. The anticancer agent rhodamine 6G, loaded into the PoPs, could be released upon exposure of the nanocomposite to the αS-binding cation Ca^{2+}. Intracellular uptake of PoPs and drug release was demonstrated with HeLa cells in the presence of intracellular Ca^{2+}-regulating agents. In this context, αS acted as a useful switch to open the gates of mesoporous SNPs by altering its conformation in a Ca^{2+}-dependent manner.

Assembly of NPs into controllable nanoobjects expands their application potential for the development of nanoscale electronic and optical devices. Biopolymers have been shown to provide effective means for organizing NPs into superstructures and fibrillar structures have attracted considerable interest in this respect. For example, pea-pod-type chains of AuNPs embedded into dielectric αS fibrils were shown to exhibit photoconductivity with visible light [141]. The unit assembly strategy of amyloid fibril formation of αS was employed to construct anisotropic one-dimensional chains of AuNPs within the amyloid fibrils. The necessary conformational transition of AuNP-adsorbed molecules from the disordered to the ordered state was induced by exposure of the assembly units to hexane or pH change. Interestingly, the morphological polymorphism of fibrillar structures obtained in different conditions could provide a means to fabricate chains of AuNPs with diverse organization. NP chains of Pd and Cu were also synthetized exploiting αS fibrils as biopolymeric templates [142]. Finally, αS amyloid fibrils were found to drive a helical arrangement of gold nanorods [143]. The latter showed no apparent interaction with the monomeric protein but effective adsorption onto chiral fibril structures via noncovalent interactions. The helical arrangement resulted in intense optical activity at the surface plasmon resonance wavelengths, thereby constituting a novel sensing technique for the detection of fibrils.

5. Conclusions

We surveyed research works focused on the behavior of the small protein αS at the interface with NPs in an attempt to obtain a more defined picture of the factors that determine the protein-particle interactions and influence the conformational transitions of the biomolecule. Through a better understanding of αS-NP systems at the molecular level, we have moved forward towards the possibility of designing nanomaterials with useful functionalities for applications in diverse scientific and technological areas. A number of studies have demonstrated that it is possible to determine and even predict how αS interacts with different nanoscale surfaces, highlighting the multiple conformations that the polypeptide can adopt and describing how to control the interactions. It has emerged that electrostatic forces dictate the mode of adsorption of monomeric αS to diverse nanoscale surfaces, providing alternative anchors (the amphiphilic N-terminal or acidic C-terminal domains) for binding to negatively or positively charged NPs. Yet, hydrophobic attraction was shown to contribute to protein adsorption too, determining some involvement of the aggregation-promoting NAC domain. It has been shown that certain NPs are able to modify protein aggregation, with encouraging results regarding the possibility of redirecting the formation of neurotoxic aggregates towards more harmless species or even of disassembling otherwise intractable amyloid fibrils. The different exposure of amino acid residues in the monomeric, oligomeric and fibrillar states of αS may provide the basis for the selective targeting of toxic species by NPs. Finally, the extraordinary conformational plasticity of αS was exploited to design higher hierarchical structures that showed interesting novel physical or chemical attributes. The field of study continues to show extraordinary vivacity and it is expected that new therapeutic and technological solutions will be proposed in the near future.

Funding: This work received no external funding.

Acknowledgments: We thank all the members of the research group for stimulating discussions.

Conflicts of Interest: The authors declare no conflict of interest.

References

1. Breydo, L.; Wu, J.W.; Uversky, V.N. α-Synuclein misfolding and Parkinson's disease. *Biochim. Biophys. Acta Mol. Basis Dis.* **2012**, *1822*, 261–285. [CrossRef] [PubMed]
2. Alderson, T.R.; Markley, J.L. Biophysical characterization of α-synuclein and its controversial structure. *Intrinsically Disord. Proteins* **2013**, *1*, e26255. [CrossRef] [PubMed]
3. Maroteaux, L.; Campanelli, J.; Scheller, R. Synuclein: A neuron-specific protein localized to the nucleus and presynaptic nerve terminal. *J. Neurosci.* **1988**, *8*, 2804–2815. [CrossRef] [PubMed]
4. Cheng, F.; Vivacqua, G.; Yu, S. The role of alpha-synuclein in neurotransmission and synaptic plasticity. *J. Chem. Neuroanat.* **2011**, *42*, 242–248. [CrossRef]
5. Burre, J.; Sharma, M.; Tsetsenis, T.; Buchman, V.; Etherton, M.R.; Sudhof, T.C. α-Synuclein Promotes SNARE-Complex Assembly in Vivo and in Vitro. *Science* **2010**, *329*, 1663–1667. [CrossRef]
6. Cabin, D.E.; Shimazu, K.; Murphy, D.; Cole, N.B.; Gottschalk, W.; McIlwain, K.L.; Orrison, B.; Chen, A.; Ellis, C.E.; Paylor, R.; et al. Synaptic Vesicle Depletion Correlates with Attenuated Synaptic Responses to Prolonged Repetitive Stimulation in Mice Lacking α-Synuclein. *J. Neurosci.* **2002**, *22*, 8797–8807. [CrossRef]
7. Burré, J.; Sharma, M.; Südhof, T.C. Cell Biology and Pathophysiology of α-Synuclein. *Cold Spring Harb. Perspect. Med.* **2018**, *8*, a024091. [CrossRef]
8. Spillantini, M.G.; Schmidt, M.L.; Lee, V.M.-Y.; Trojanowski, J.Q.; Jakes, R.; Goedert, M. α-Synuclein in Lewy bodies. *Nature* **1997**, *388*, 839–840. [CrossRef]
9. Uéda, K.; Fukushima, H.; Masliah, E.; Xia, Y.; Iwai, A.; Yoshimoto, M.; Otero, D.A.; Kondo, J.; Ihara, Y.; Saitoh, T. Molecular cloning of cDNA encoding an unrecognized component of amyloid in Alzheimer disease. *Proc. Natl. Acad. Sci. USA* **1993**, *90*, 11282–11286. [CrossRef]
10. Lashuel, H.A.; Overk, C.R.; Oueslati, A.; Masliah, E. The many faces of α-synuclein: From structure and toxicity to therapeutic target. *Nat. Rev. Neurosci.* **2013**, *14*, 38–48. [CrossRef]
11. Weinreb, P.H.; Zhen, W.; Poon, A.W.; Conway, K.A.; Lansbury, P.T. NACP, A Protein Implicated in Alzheimer's Disease and Learning, Is Natively Unfolded. *Biochemistry* **1996**, *35*, 13709–13715. [CrossRef] [PubMed]
12. Eliezer, D.; Kutluay, E.; Bussell, R.; Browne, G. Conformational properties of α-synuclein in its free and lipid-associated states. *J. Mol. Biol.* **2001**, *307*, 1061–1073. [CrossRef] [PubMed]
13. Nath, A.; Sammalkorpi, M.; DeWitt, D.C.; Trexler, A.J.; Elbaum-Garfinkle, S.; O'Hern, C.S.; Rhoades, E. The Conformational Ensembles of α-Synuclein and Tau: Combining Single-Molecule FRET and Simulations. *Biophys. J.* **2012**, *103*, 1940–1949. [CrossRef] [PubMed]
14. Ullman, O.; Fisher, C.K.; Stultz, C.M. Explaining the Structural Plasticity of α-Synuclein. *J. Am. Chem. Soc.* **2011**, *133*, 19536–19546. [CrossRef]
15. Bartels, T.; Choi, J.G.; Selkoe, D.J. α-Synuclein occurs physiologically as a helically folded tetramer that resists aggregation. *Nature* **2011**, *477*, 107–110. [CrossRef]
16. Wang, W.; Perovic, I.; Chittuluru, J.; Kaganovich, A.; Nguyen, L.T.T.; Liao, J.; Auclair, J.R.; Johnson, D.; Landeru, A.; Simorellis, A.K.; et al. A soluble α-synuclein construct forms a dynamic tetramer. *Proc. Natl. Acad. Sci. USA* **2011**, *108*, 17797–17802. [CrossRef]
17. Dettmer, U.; Newman, A.J.; Luth, E.S.; Bartels, T.; Selkoe, D. In Vivo Cross-linking Reveals Principally Oligomeric Forms of α-Synuclein and β-Synuclein in Neurons and Non-neural Cells. *J. Biol. Chem.* **2013**, *288*, 6371–6385. [CrossRef]
18. Gould, N.; Mor, D.E.; Lightfoot, R.; Malkus, K.; Giasson, B.; Ischiropoulos, H. Evidence of Native α-Synuclein Conformers in the Human Brain. *J. Biol. Chem.* **2014**, *289*, 7929–7934. [CrossRef]
19. Theillet, F.-X.; Binolfi, A.; Bekei, B.; Martorana, A.; Rose, H.M.; Stuiver, M.; Verzini, S.; Lorenz, D.; van Rossum, M.; Goldfarb, D.; et al. Structural disorder of monomeric α-synuclein persists in mammalian cells. *Nature* **2016**, *530*, 45–50. [CrossRef]

20. Guerrero-Ferreira, R.; Taylor, N.M.; Mona, D.; Ringler, P.; Lauer, M.E.; Riek, R.; Britschgi, M.; Stahlberg, H. Cryo-EM structure of alpha-synuclein fibrils. *eLife* **2018**, *7*, e36402. [CrossRef]

21. Giehm, L.; Lorenzen, N.; Otzen, D.E. Assays for α-synuclein aggregation. *Methods* **2011**, *53*, 295–305. [CrossRef] [PubMed]

22. Mahmoudi, M.; Kalhor, H.R.; Laurent, S.; Lynch, I. Protein fibrillation and nanoparticle interactions: Opportunities and challenges. *Nanoscale* **2013**, *5*, 2570. [CrossRef] [PubMed]

23. Nel, A.E.; Mädler, L.; Velegol, D.; Xia, T.; Hoek, E.M.V.; Somasundaran, P.; Klaessig, F.; Castranova, V.; Thompson, M. Understanding biophysicochemical interactions at the nano–bio interface. *Nat. Mater* **2009**, *8*, 543–557. [CrossRef] [PubMed]

24. Verma, A.; Rotello, V.M. Surface recognition of biomacromolecules using nanoparticle receptors. *Chem. Commun.* **2005**, 303–312. [CrossRef] [PubMed]

25. Stark, W.J. Nanoparticles in Biological Systems. *Angew. Chem. Int. Ed.* **2011**, *50*, 1242–1258. [CrossRef] [PubMed]

26. Zanzoni, S.; Ceccon, A.; Assfalg, M.; Singh, R.K.; Fushman, D.; D'Onofrio, M. Polyhydroxylated [60]fullerene binds specifically to functional recognition sites on a monomeric and a dimeric ubiquitin. *Nanoscale* **2015**, *7*, 7197–7205. [CrossRef] [PubMed]

27. Bortot, A.; Zanzoni, S.; D'Onofrio, M.; Assfalg, M. Specific Interaction Sites Determine Differential Adsorption of Protein Structural Isomers on Nanoparticle Surfaces. *Chem. A Eur. J.* **2018**, *24*, 5911–5919. [CrossRef] [PubMed]

28. Roach, P.; Farrar, D.; Perry, C.C. Surface Tailoring for Controlled Protein Adsorption: Effect of Topography at the Nanometer Scale and Chemistry. *J. Am. Chem. Soc.* **2006**, *128*, 3939–3945. [CrossRef]

29. Davis, M.E.; Chen, Z.; Shin, D.M. Nanoparticle therapeutics: An emerging treatment modality for cancer. *Nat. Rev. Drug Discov.* **2008**, *7*, 771–782. [CrossRef]

30. Monopoli, M.P.; Åberg, C.; Salvati, A.; Dawson, K.A. Biomolecular coronas provide the biological identity of nanosized materials. *Nat. Nanotechnol.* **2012**, *7*, 779–786. [CrossRef]

31. Fleischer, C.C.; Payne, C.K. Nanoparticle–Cell Interactions: Molecular Structure of the Protein Corona and Cellular Outcomes. *Acc. Chem. Res.* **2014**, *47*, 2651–2659. [CrossRef] [PubMed]

32. Roach, P.; Farrar, D.; Perry, C.C. Interpretation of Protein Adsorption: Surface-Induced Conformational Changes. *J. Am. Chem. Soc.* **2005**, *127*, 8168–8173. [CrossRef] [PubMed]

33. Lynch, I.; Dawson, K.A. Protein-nanoparticle interactions. *Nano Today* **2008**, *3*, 40–47. [CrossRef]

34. Rana, S.; Yeh, Y.-C.; Rotello, V.M. Engineering the nanoparticle–protein interface: Applications and possibilities. *Curr. Opin. Chem. Biol.* **2010**, *14*, 828–834. [CrossRef]

35. Sapsford, K.E.; Tyner, K.M.; Dair, B.J.; Deschamps, J.R.; Medintz, I.L. Analyzing Nanomaterial Bioconjugates: A Review of Current and Emerging Purification and Characterization Techniques. *Anal. Chem.* **2011**, *83*, 4453–4488. [CrossRef]

36. Yang, J.A.; Lin, W.; Woods, W.S.; George, J.M.; Murphy, C.J. α-Synuclein's Adsorption, Conformation and Orientation on Cationic Gold Nanoparticle Surfaces Seeds Global Conformation Change. *J. Phys. Chem. B* **2014**, *118*, 3559–3571. [CrossRef]

37. Vitali, M.; Rigamonti, V.; Natalello, A.; Colzani, B.; Avvakumova, S.; Brocca, S.; Santambrogio, C.; Narkiewicz, J.; Legname, G.; Colombo, M.; et al. Conformational properties of intrinsically disordered proteins bound to the surface of silica nanoparticles. *Biochim. Biophys. Acta Gen. Subj.* **2018**, *1862*, 1556–1564. [CrossRef]

38. Tira, R.; De Cecco, E.; Rigamonti, V.; Santambrogio, C.; Barracchia, C.G.; Munari, F.; Romeo, A.; Legname, G.; Prosperi, D.; Grandori, R.; et al. Dynamic molecular exchange and conformational transitions of alpha-synuclein at the nano-bio interface. *Int. J. Biol. Macromol.* **2020**, *154*, 206–216. [CrossRef]

39. Galvagnion, C.; Buell, A.K.; Meisl, G.; Michaels, T.C.T.; Vendruscolo, M.; Knowles, T.P.J.; Dobson, C.M. Lipid vesicles trigger α-synuclein aggregation by stimulating primary nucleation. *Nat. Chem. Biol.* **2015**, *11*, 229–234. [CrossRef]

40. Kim, D.; Yoo, J.M.; Hwang, H.; Lee, J.; Lee, S.H.; Yun, S.P.; Park, M.J.; Lee, M.; Choi, S.; Kwon, S.H.; et al. Graphene quantum dots prevent α-synucleinopathy in Parkinson's disease. *Nat. Nanotechnol.* **2018**, *13*, 812–818. [CrossRef]

41. Álvarez, Y.D.; Fauerbach, J.A.; Pellegrotti, J.V.; Jovin, T.M.; Jares-Erijman, E.A.; Stefani, F.D. Influence of Gold Nanoparticles on the Kinetics of α-Synuclein Aggregation. *Nano Lett.* **2013**, *13*, 6156–6163. [CrossRef] [PubMed]

42. Vácha, R.; Linse, S.; Lund, M. Surface Effects on Aggregation Kinetics of Amyloidogenic Peptides. *J. Am. Chem. Soc.* **2014**, *136*, 11776–11782. [CrossRef] [PubMed]

43. Joshi, N.; Basak, S.; Kundu, S.; De, G.; Mukhopadhyay, A.; Chattopadhyay, K. Attenuation of the Early Events of α-Synuclein Aggregation: A Fluorescence Correlation Spectroscopy and Laser Scanning Microscopy Study in the Presence of Surface-Coated Fe_3O_4 Nanoparticles. *Langmuir* **2015**, *31*, 1469–1478. [CrossRef] [PubMed]

44. Mohammad-Beigi, H.; Hosseini, A.; Adeli, M.; Ejtehadi, M.R.; Christiansen, G.; Sahin, C.; Tu, Z.; Tavakol, M.; Dilmaghani-Marand, A.; Nabipour, I.; et al. Mechanistic Understanding of the Interactions between Nano-Objects with Different Surface Properties and α-Synuclein. *ACS Nano* **2019**, *13*, 3243–3256. [CrossRef]

45. Roberti, M.J.; Morgan, M.; Menéndez, G.; Pietrasanta, L.I.; Jovin, T.M.; Jares-Erijman, E.A. Quantum Dots As Ultrasensitive Nanoactuators and Sensors of Amyloid Aggregation in Live Cells. *J. Am. Chem. Soc.* **2009**, *131*, 8102–8107. [CrossRef]

46. Lee, J.; Bhak, G.; Lee, J.-H.; Park, W.; Lee, M.; Lee, D.; Jeon, N.L.; Jeong, D.H.; Char, K.; Paik, S.R. Free-Standing Gold-Nanoparticle Monolayer Film Fabricated by Protein Self-Assembly of α-Synuclein. *Angew. Chem. Int. Ed.* **2015**, *54*, 4571–4576. [CrossRef]

47. Assfalg, M.; Ragona, L.; Pagano, K.; D'Onofrio, M.; Zanzoni, S.; Tomaselli, S.; Molinari, H. The study of transient protein–nanoparticle interactions by solution NMR spectroscopy. *Biochim. Biophys. Acta Proteins Proteom.* **2016**, *1864*, 102–114. [CrossRef]

48. Randika Perera, Y.; Hill, R.A.; Fitzkee, N.C. Protein Interactions with Nanoparticle Surfaces: Highlighting Solution NMR Techniques. *ISR J. Chem.* **2019**, *59*, 962–979. [CrossRef]

49. Xie, M.; Hansen, A.L.; Yuan, J.; Brüschweiler, R. Residue-Specific Interactions of an Intrinsically Disordered Protein with Silica Nanoparticles and Their Quantitative Prediction. *J. Phys. Chem. C* **2016**, *120*, 24463–24468. [CrossRef]

50. Xie, M.; Li, D.; Yuan, J.; Hansen, A.L.; Brüschweiler, R. Quantitative Binding Behavior of Intrinsically Disordered Proteins to Nanoparticle Surfaces at Individual Residue Level. *Chem. A Eur. J.* **2018**, *24*, 16997–17001. [CrossRef]

51. Li, D.-W.; Xie, M.; Brüschweiler, R. Quantitative Cooperative Binding Model for Intrinsically Disordered Proteins Interacting with Nanomaterials. *J. Am. Chem. Soc.* **2020**, *142*, 10730–10738. [CrossRef] [PubMed]

52. Henzler, K.; Haupt, B.; Rosenfeldt, S.; Harnau, L.; Narayanan, T.; Ballauff, M. Interaction strength between proteins and polyelectrolyte brushes: A small angle X-ray scattering study. *Phys. Chem. Chem. Phys.* **2011**, *13*, 17599. [CrossRef] [PubMed]

53. Clemments, A.M.; Botella, P.; Landry, C.C. Spatial Mapping of Protein Adsorption on Mesoporous Silica Nanoparticles by Stochastic Optical Reconstruction Microscopy. *J. Am. Chem. Soc.* **2017**, *139*, 3978–3981. [CrossRef] [PubMed]

54. Shang, L.; Nienhaus, G.U. In Situ Characterization of Protein Adsorption onto Nanoparticles by Fluorescence Correlation Spectroscopy. *Acc. Chem. Res.* **2017**, *50*, 387–395. [CrossRef]

55. Yang, J.A.; Johnson, B.J.; Wu, S.; Woods, W.S.; George, J.M.; Murphy, C.J. Study of Wild-Type α-Synuclein Binding and Orientation on Gold Nanoparticles. *Langmuir* **2013**, *29*, 4603–4615. [CrossRef]

56. Lin, W.; Insley, T.; Tuttle, M.D.; Zhu, L.; Berthold, D.A.; Král, P.; Rienstra, C.M.; Murphy, C.J. Control of Protein Orientation on Gold Nanoparticles. *J. Phys. Chem. C* **2015**, *119*, 21035–21043. [CrossRef]

57. Iwai, A.; Masliah, E.; Yoshimoto, M.; Ge, N.; Flanagan, L.; Rohan de Silva, H.A.; Kittel, A.; Saitoh, T. The precursor protein of non-Aβ component of Alzheimer's disease amyloid is a presynaptic protein of the central nervous system. *Neuron* **1995**, *14*, 467–475. [CrossRef]

58. Kahle, P.J.; Neumann, M.; Ozmen, L.; Müller, V.; Jacobsen, H.; Schindzielorz, A.; Okochi, M.; Leimer, U.; van der Putten, H.; Probst, A.; et al. Subcellular Localization of Wild-Type and Parkinson's Disease-Associated Mutant α-Synuclein in Human and Transgenic Mouse Brain. *J. Neurosci.* **2000**, *20*, 6365–6373. [CrossRef]

59. Jensen, P.H.; Nielsen, M.S.; Jakes, R.; Dotti, C.G.; Goedert, M. Binding of α-Synuclein to Brain Vesicles Is Abolished by Familial Parkinson's Disease Mutation. *J. Biol. Chem.* **1998**, *273*, 26292–26294. [CrossRef]

60. Perez, R.G.; Waymire, J.C.; Lin, E.; Liu, J.J.; Guo, F.; Zigmond, M.J. A Role for α-Synuclein in the Regulation of Dopamine Biosynthesis. *J. Neurosci.* **2002**, *22*, 3090–3099. [CrossRef]

61. Davidson, W.S.; Jonas, A.; Clayton, D.F.; George, J.M. Stabilization of α-Synuclein Secondary Structure upon Binding to Synthetic Membranes. *J. Biol. Chem.* **1998**, *273*, 9443–9449. [CrossRef] [PubMed]

62. Perrin, R.J.; Woods, W.S.; Clayton, D.F.; George, J.M. Interaction of Human α-Synuclein and Parkinson's Disease Variants with Phospholipids: Structural analysis using site-directed mutagenesis. *J. Biol. Chem.* **2000**, *275*, 34393–34398. [CrossRef]

63. Chandra, S.; Chen, X.; Rizo, J.; Jahn, R.; Südhof, T.C. A Broken α-Helix in Folded α-Synuclein. *J. Biol. Chem.* **2003**, *278*, 15313–15318. [CrossRef] [PubMed]

64. Beyer, K. Mechanistic aspects of Parkinson's disease: α-synuclein and the biomembrane. *Cell Biochem. Biophys.* **2007**, *47*, 285–299. [CrossRef] [PubMed]

65. Bussell, R.; Eliezer, D. A Structural and Functional Role for 11-mer Repeats in α-Synuclein and Other Exchangeable Lipid Binding Proteins. *J. Mol. Biol.* **2003**, *329*, 763–778. [CrossRef]

66. Bisaglia, M.; Tessari, I.; Pinato, L.; Bellanda, M.; Giraudo, S.; Fasano, M.; Bergantino, E.; Bubacco, L.; Mammi, S. A Topological Model of the Interaction between α-Synuclein and Sodium Dodecyl Sulfate Micelles. *Biochemistry* **2005**, *44*, 329–339. [CrossRef]

67. Ulmer, T.S.; Bax, A.; Cole, N.B.; Nussbaum, R.L. Structure and Dynamics of Micelle-bound Human-Synuclein. *J. Biol. Chem.* **2005**, *280*, 9595–9603. [CrossRef]

68. Jao, C.C.; Hegde, B.G.; Chen, J.; Haworth, I.S.; Langen, R. Structure of membrane-bound α-synuclein from site-directed spin labeling and computational refinement. *Proc. Natl. Acad. Sci. USA* **2008**, *105*, 19666–19671. [CrossRef]

69. Trexler, A.J.; Rhoades, E. α-Synuclein Binds Large Unilamellar Vesicles as an Extended Helix. *Biochemistry* **2009**, *48*, 2304–2306. [CrossRef]

70. Drescher, M.; Veldhuis, G.; van Rooijen, B.D.; Milikisyants, S.; Subramaniam, V.; Huber, M. Antiparallel Arrangement of the Helices of Vesicle-Bound α-Synuclein. *J. Am. Chem. Soc.* **2008**, *130*, 7796–7797. [CrossRef]

71. Ferreon, A.C.M.; Gambin, Y.; Lemke, E.A.; Deniz, A.A. Interplay of α-synuclein binding and conformational switching probed by single-molecule fluorescence. *Proc. Natl. Acad. Sci. USA* **2009**, *106*, 5645–5650. [CrossRef] [PubMed]

72. Georgieva, E.R.; Ramlall, T.F.; Borbat, P.P.; Freed, J.H.; Eliezer, D. The Lipid-binding Domain of Wild Type and Mutant α-Synuclein: Compactness and interconversion between the broken and extended helix forms. *J. Biol. Chem.* **2010**, *285*, 28261–28274. [CrossRef] [PubMed]

73. Bodner, C.R.; Dobson, C.M.; Bax, A. Multiple Tight Phospholipid-Binding Modes of α-Synuclein Revealed by Solution NMR Spectroscopy. *J. Mol. Biol.* **2009**, *390*, 775–790. [CrossRef] [PubMed]

74. Viennet, T.; Wördehoff, M.M.; Uluca, B.; Poojari, C.; Shaykhalishahi, H.; Willbold, D.; Strodel, B.; Heise, H.; Buell, A.K.; Hoyer, W.; et al. Structural insights from lipid-bilayer nanodiscs link α-Synuclein membrane-binding modes to amyloid fibril formation. *Commun. Biol.* **2018**, *1*, 44. [CrossRef]

75. Rhoades, E.; Ramlall, T.F.; Webb, W.W.; Eliezer, D. Quantification of α-Synuclein Binding to Lipid Vesicles Using Fluorescence Correlation Spectroscopy. *Biophys. J.* **2006**, *90*, 4692–4700. [CrossRef]

76. Kubo, S.; Nemani, V.M.; Chalkley, R.J.; Anthony, M.D.; Hattori, N.; Mizuno, Y.; Edwards, R.H.; Fortin, D.L. A Combinatorial Code for the Interaction of α-Synuclein with Membranes. *J. Biol. Chem.* **2005**, *280*, 31664–31672. [CrossRef]

77. Eichmann, C.; Campioni, S.; Kowal, J.; Maslennikov, I.; Gerez, J.; Liu, X.; Verasdonck, J.; Nespovitaya, N.; Choe, S.; Meier, B.H.; et al. Preparation and Characterization of Stable α-Synuclein Lipoprotein Particles. *J. Biol. Chem.* **2016**, *291*, 8516–8527. [CrossRef]

78. Chung, P.J.; Zhang, Q.; Hwang, H.L.; Leong, A.; Maj, P.; Szczygiel, R.; Dufresne, E.M.; Narayanan, S.; Adams, E.J.; Lee, K.Y.C. α-Synuclein Sterically Stabilizes Spherical Nanoparticle-Supported Lipid Bilayers. *ACS Appl. Bio Mater.* **2019**, *2*, 1413–1419. [CrossRef]

79. McClain, S.M.; Ojoawo, A.M.; Lin, W.; Rienstra, C.M.; Murphy, C.J. Interaction of Alpha-Synuclein and Its Mutants with Rigid Lipid Vesicle Mimics of Varying Surface Curvature. *ACS Nano* **2020**, *14*, 10153–10167. [CrossRef]

80. Mohammad-Beigi, H.; Shojaosadati, S.A.; Marvian, A.T.; Pedersen, J.N.; Klausen, L.H.; Christiansen, G.; Pedersen, J.S.; Dong, M.; Morshedi, D.; Otzen, D.E. Strong interactions with polyethylenimine-coated human serum albumin nanoparticles (PEI-HSA NPs) alter α-synuclein conformation and aggregation kinetics. *Nanoscale* **2015**, *7*, 19627–19640. [CrossRef]

81. Cremades, N.; Cohen, S.I.A.; Deas, E.; Abramov, A.Y.; Chen, A.Y.; Orte, A.; Sandal, M.; Clarke, R.W.; Dunne, P.; Aprile, F.A.; et al. Direct Observation of the Interconversion of Normal and Toxic Forms of α-Synuclein. *Cell* **2012**, *149*, 1048–1059. [CrossRef] [PubMed]

82. Iljina, M.; Garcia, G.A.; Horrocks, M.H.; Tosatto, L.; Choi, M.L.; Ganzinger, K.A.; Abramov, A.Y.; Gandhi, S.; Wood, N.W.; Cremades, N.; et al. Kinetic model of the aggregation of alpha-synuclein provides insights into prion-like spreading. *Proc. Natl. Acad. Sci. USA* **2016**, *113*, E1206–E1215. [CrossRef] [PubMed]

83. Cohen, S.I.A.; Vendruscolo, M.; Welland, M.E.; Dobson, C.M.; Terentjev, E.M.; Knowles, T.P.J. Nucleated polymerization with secondary pathways. I. Time evolution of the principal moments. *J. Chem. Phys.* **2011**, *135*, 065105. [CrossRef] [PubMed]

84. Cohen, S.I.A.; Vendruscolo, M.; Dobson, C.M.; Knowles, T.P.J. From Macroscopic Measurements to Microscopic Mechanisms of Protein Aggregation. *J. Mol. Biol.* **2012**, *421*, 160–171. [CrossRef] [PubMed]

85. Knowles, T.P.J.; Waudby, C.A.; Devlin, G.L.; Cohen, S.I.A.; Aguzzi, A.; Vendruscolo, M.; Terentjev, E.M.; Welland, M.E.; Dobson, C.M. An Analytical Solution to the Kinetics of Breakable Filament Assembly. *Science* **2009**, *326*, 1533–1537. [CrossRef] [PubMed]

86. Fauerbach, J.A.; Yushchenko, D.A.; Shahmoradian, S.H.; Chiu, W.; Jovin, T.M.; Jares-Erijman, E.A. Supramolecular Non-Amyloid Intermediates in the Early Stages of α-Synuclein Aggregation. *Biophys. J.* **2012**, *102*, 1127–1136. [CrossRef] [PubMed]

87. Rabe, M.; Soragni, A.; Reynolds, N.P.; Verdes, D.; Liverani, E.; Riek, R.; Seeger, S. On-Surface Aggregation of α-Synuclein at Nanomolar Concentrations Results in Two Distinct Growth Mechanisms. *ACS Chem. Neurosci.* **2013**, *4*, 408–417. [CrossRef]

88. Aisenbrey, C.; Borowik, T.; Byström, R.; Bokvist, M.; Lindström, F.; Misiak, H.; Sani, M.-A.; Gröbner, G. How is protein aggregation in amyloidogenic diseases modulated by biological membranes? *Eur. Biophys. J.* **2008**, *37*, 247–255. [CrossRef]

89. Relini, A.; Cavalleri, O.; Rolandi, R.; Gliozzi, A. The two-fold aspect of the interplay of amyloidogenic proteins with lipid membranes. *Chem. Phys. Lipids* **2009**, *158*, 1–9. [CrossRef]

90. Selkoe, D.J. Cell biology of protein misfolding: The examples of Alzheimer's and Parkinson's diseases. *Nat. Cell Biol.* **2004**, *6*, 1054–1061. [CrossRef]

91. Butterfield, S.M.; Lashuel, H.A. Amyloidogenic Protein-Membrane Interactions: Mechanistic Insight from Model Systems. *Angew. Chem. Int. Ed.* **2010**, *49*, 5628–5654. [CrossRef] [PubMed]

92. Hebda, J.A.; Miranker, A.D. The Interplay of Catalysis and Toxicity by Amyloid Intermediates on Lipid Bilayers: Insights from Type II Diabetes. *Annu. Rev. Biophys.* **2009**, *38*, 125–152. [CrossRef] [PubMed]

93. Byström, R.; Aisenbrey, C.; Borowik, T.; Bokvist, M.; Lindström, F.; Sani, M.-A.; Olofsson, A.; Gröbner, G. Disordered Proteins: Biological Membranes as Two-Dimensional Aggregation Matrices. *Cell Biochem. Biophys.* **2008**, *52*, 175–189. [CrossRef] [PubMed]

94. Linse, S.; Cabaleiro-Lago, C.; Xue, W.-F.; Lynch, I.; Lindman, S.; Thulin, E.; Radford, S.E.; Dawson, K.A. Nucleation of protein fibrillation by nanoparticles. *Proc. Natl. Acad. Sci. USA* **2007**, *104*, 8691–8696. [CrossRef] [PubMed]

95. Sauvage, F.; Schymkowitz, J.; Rousseau, F.; Schmidt, B.Z.; Remaut, K.; Braeckmans, K.; De Smedt, S.C. Nanomaterials to avoid and destroy protein aggregates. *Nano Today* **2020**, *31*, 100837. [CrossRef]

96. Lombardo, S.M.; Schneider, M.; Türeli, A.E.; Günday Türeli, N. Key for crossing the BBB with nanoparticles: The rational design. *Beilstein J. Nanotechnol.* **2020**, *11*, 866–883. [CrossRef]

97. Behzadi, S.; Serpooshan, V.; Tao, W.; Hamaly, M.A.; Alkawareek, M.Y.; Dreaden, E.C.; Brown, D.; Alkilany, A.M.; Farokhzad, O.C.; Mahmoudi, M. Cellular uptake of nanoparticles: Journey inside the cell. *Chem. Soc. Rev.* **2017**, *46*, 4218–4244. [CrossRef]

98. Donahue, N.D.; Acar, H.; Wilhelm, S. Concepts of nanoparticle cellular uptake, intracellular trafficking and kinetics in nanomedicine. *Adv. Drug Deliv. Rev.* **2019**, *143*, 68–96. [CrossRef]

99. Tahaei Gilan, S.S.; Yahya Rayat, D.; Ahmed Mustafa, T.; Aziz, F.M.; Shahpasand, K.; Akhtari, K.; Salihi, A.; Abou-Zied, O.K.; Falahati, M. α-synuclein interaction with zero-valent iron nanoparticles accelerates structural rearrangement into amyloid-susceptible structure with increased cytotoxic tendency. *Int. J. Nanomed.* **2019**, *14*, 4637–4648. [CrossRef]

100. Gao, G.; Chen, R.; He, M.; Li, J.; Li, J.; Wang, L.; Sun, T. Gold nanoclusters for Parkinson's disease treatment. *Biomaterials* **2019**, *194*, 36–46. [CrossRef]

101. Asthana, S.; Bhattacharyya, D.; Kumari, S.; Nayak, P.S.; Saleem, M.; Bhunia, A.; Jha, S. Interaction with zinc oxide nanoparticle kinetically traps α-synuclein fibrillation into off-pathway non-toxic intermediates. *Int. J. Biol. Macromol.* **2020**, *150*, 68–79. [CrossRef] [PubMed]

102. Zand, Z.; Afarinesh Khaki, P.; Salihi, A.; Sharifi, M.; Qadir Nanakali, N.M.; Alasady, A.A.; Mohammad Aziz, F.; Shahpasand, K.; Hasan, A.; Falahati, M. Cerium oxide NPs mitigate the amyloid formation of α-synuclein and associated cytotoxicity. *Int. J. Nanomed.* **2019**, *14*, 6989–7000. [CrossRef] [PubMed]

103. Ruotolo, R.; De Giorgio, G.; Minato, I.; Bianchi, M.; Bussolati, O.; Marmiroli, N. Cerium Oxide Nanoparticles Rescue α-Synuclein-Induced Toxicity in a Yeast Model of Parkinson's Disease. *Nanomaterials* **2020**, *10*, 235. [CrossRef] [PubMed]

104. Taebnia, N.; Morshedi, D.; Doostkam, M.; Yaghmaei, S.; Aliakbari, F.; Singh, G.; Arpanaei, A. The effect of mesoporous silica nanoparticle surface chemistry and concentration on the α-synuclein fibrillation. *RSC Adv.* **2015**, *5*, 60966–60974. [CrossRef]

105. Mohammadi, S.; Nikkhah, M.; Hosseinkhani, S. Investigation of the effects of carbon-based nanomaterials on A53T alpha-synuclein aggregation using a whole-cell recombinant biosensor. *Int. J. Nanomed.* **2017**, *12*, 8831–8840. [CrossRef] [PubMed]

106. Sun, Y.; Kakinen, A.; Zhang, C.; Yang, Y.; Faridi, A.; Davis, T.P.; Cao, W.; Ke, P.C.; Ding, F. Amphiphilic surface chemistry of fullerenols is necessary for inhibiting the amyloid aggregation of alpha-synuclein NACore. *Nanoscale* **2019**, *11*, 11933–11945. [CrossRef] [PubMed]

107. Milowska, K.; Grochowina, J.; Katir, N.; El Kadib, A.; Majoral, J.-P.; Bryszewska, M.; Gabryelak, T. Viologen-Phosphorus Dendrimers Inhibit α-Synuclein Fibrillation. *Mol. Pharm.* **2013**, *10*, 1131–1137. [CrossRef] [PubMed]

108. Milowska, K.; Malachowska, M.; Gabryelak, T. PAMAM G4 dendrimers affect the aggregation of α-synuclein. *Int. J. Biol. Macromol.* **2011**, *48*, 742–746. [CrossRef]

109. Rekas, A.; Lo, V.; Gadd, G.E.; Cappai, R.; Yun, S.I. PAMAM Dendrimers as Potential Agents against Fibrillation of α -Synuclein, a Parkinson's Disease-Related Protein. *Macromol. Biosci.* **2009**, *9*, 230–238. [CrossRef]

110. Milowska, K.; Gabryelak, T.; Bryszewska, M.; Caminade, A.-M.; Majoral, J.-P. Phosphorus-containing dendrimers against α-synuclein fibril formation. *Int. J. Biol. Macromol.* **2012**, *50*, 1138–1143. [CrossRef]

111. Milowska, K.; Szwed, A.; Mutrynowska, M.; Gomez-Ramirez, R.; de la Mata, F.J.; Gabryelak, T.; Bryszewska, M. Carbosilane dendrimers inhibit α-synuclein fibrillation and prevent cells from rotenone-induced damage. *Int. J. Pharm.* **2015**, *484*, 268–275. [CrossRef] [PubMed]

112. Mohammad-Beigi, H.; Morshedi, D.; Shojaosadati, S.A.; Pedersen, J.N.; Marvian, A.T.; Aliakbari, F.; Christiansen, G.; Pedersen, J.S.; Otzen, D.E. Gallic acid loaded onto polyethylenimine-coated human serum albumin nanoparticles (PEI-HSA-GA NPs) stabilizes α-synuclein in the unfolded conformation and inhibits aggregation. *RSC Adv.* **2016**, *6*, 85312–85323. [CrossRef]

113. Aliakbari, F.; Mohammad-Beigi, H.; Rezaei-Ghaleh, N.; Becker, S.; Dehghani Esmatabad, F.; Eslampanah Seyedi, H.A.; Bardania, H.; Tayaranian Marvian, A.; Collingwood, J.F.; Christiansen, G.; et al. The potential of zwitterionic nanoliposomes against neurotoxic alpha-synuclein aggregates in Parkinson's Disease. *Nanoscale* **2018**, *10*, 9174–9185. [CrossRef] [PubMed]

114. Zhu, M.; Li, J.; Fink, A.L. The Association of α-Synuclein with Membranes Affects Bilayer Structure, Stability and Fibril Formation. *J. Biol. Chem.* **2003**, *278*, 40186–40197. [CrossRef] [PubMed]

115. Zhu, M.; Fink, A.L. Lipid Binding Inhibits α-Synuclein Fibril Formation. *J. Biol. Chem.* **2003**, *278*, 16873–16877. [CrossRef] [PubMed]

116. Yushchenko, D.A.; Fauerbach, J.A.; Thirunavukkuarasu, S.; Jares-Erijman, E.A.; Jovin, T.M. Fluorescent Ratiometric MFC Probe Sensitive to Early Stages of α-Synuclein Aggregation. *J. Am. Chem. Soc.* **2010**, *132*, 7860–7861. [CrossRef] [PubMed]

117. Meisl, G.; Kirkegaard, J.B.; Arosio, P.; Michaels, T.C.T.; Vendruscolo, M.; Dobson, C.M.; Linse, S.; Knowles, T.P.J. Molecular mechanisms of protein aggregation from global fitting of kinetic models. *Nat. Protoc.* **2016**, *11*, 252–272. [CrossRef]

118. Brancolini, G.; Corazza, A.; Vuano, M.; Fogolari, F.; Mimmi, M.C.; Bellotti, V.; Stoppini, M.; Corni, S.; Esposito, G. Probing the Influence of Citrate-Capped Gold Nanoparticles on an Amyloidogenic Protein. *ACS Nano* **2015**, *9*, 2600–2613. [CrossRef]

119. Zhou, W.; Long, C.; Reaney, S.H.; Di Monte, D.A.; Fink, A.L.; Uversky, V.N. Methionine oxidation stabilizes non-toxic oligomers of α-synuclein through strengthening the auto-inhibitory intra-molecular long-range interactions. *Biochim. Biophys. Acta Mol. Basis Dis.* **2010**, *1802*, 322–330. [CrossRef]

120. Van Meer, G.; Voelker, D.R.; Feigenson, G.W. Membrane lipids: Where they are and how they behave. *Nat. Rev. Mol. Cell Biol.* **2008**, *9*, 112–124. [CrossRef]

121. Vanni, S.; Hirose, H.; Barelli, H.; Antonny, B.; Gautier, R. A sub-nanometre view of how membrane curvature and composition modulate lipid packing and protein recruitment. *Nat. Commun.* **2014**, *5*, 4916. [CrossRef] [PubMed]

122. Kjaer, L.; Giehm, L.; Heimburg, T.; Otzen, D. The Influence of Vesicle Size and Composition on α-Synuclein Structure and Stability. *Biophys. J.* **2009**, *96*, 2857–2870. [CrossRef]

123. Galvagnion, C.; Brown, J.W.P.; Ouberai, M.M.; Flagmeier, P.; Vendruscolo, M.; Buell, A.K.; Sparr, E.; Dobson, C.M. Chemical properties of lipids strongly affect the kinetics of the membrane-induced aggregation of α-synuclein. *Proc. Natl. Acad. Sci. USA* **2016**, *113*, 7065–7070. [CrossRef] [PubMed]

124. Fink, A.L. The Aggregation and Fibrillation of α-Synuclein. *Acc. Chem. Res.* **2006**, *39*, 628–634. [CrossRef] [PubMed]

125. Giehm, L.; Svergun, D.I.; Otzen, D.E.; Vestergaard, B. Low-resolution structure of a vesicle disrupting α-synuclein oligomer that accumulates during fibrillation. *Proc. Natl. Acad. Sci. USA* **2011**, *108*, 3246–3251. [CrossRef]

126. Hellstrand, E.; Nowacka, A.; Topgaard, D.; Linse, S.; Sparr, E. Membrane Lipid Co-Aggregation with α-Synuclein Fibrils. *PLoS ONE* **2013**, *8*, e77235. [CrossRef]

127. Barracchia, C.G.; Tira, R.; Parolini, F.; Munari, F.; Bubacco, L.; Spyroulias, G.A.; D'Onofrio, M.; Assfalg, M. Unsaturated Fatty Acid-Induced Conformational Transitions and Aggregation of the Repeat Domain of Tau. *Molecules* **2020**, *25*, 2716. [CrossRef]

128. Sparr, E.; Linse, S. Lipid-protein interactions in amyloid formation. *Biochim. Biophys. Acta Proteins Proteom.* **2019**, *1867*, 455–457. [CrossRef]

129. Auluck, P.K.; Caraveo, G.; Lindquist, S. α-Synuclein: Membrane Interactions and Toxicity in Parkinson's Disease. *Annu. Rev. Cell Dev. Biol.* **2010**, *26*, 211–233. [CrossRef]

130. Nemani, V.M.; Lu, W.; Berge, V.; Nakamura, K.; Onoa, B.; Lee, M.K.; Chaudhry, F.A.; Nicoll, R.A.; Edwards, R.H. Increased Expression of α-Synuclein Reduces Neurotransmitter Release by Inhibiting Synaptic Vesicle Reclustering after Endocytosis. *Neuron* **2010**, *65*, 66–79. [CrossRef]

131. Rovere, M.; Sanderson, J.B.; Fonseca-Ornelas, L.; Patel, D.S.; Bartels, T. Refolding of helical soluble α-synuclein through transient interaction with lipid interfaces. *FEBS Lett.* **2018**, *592*, 1464–1472. [CrossRef] [PubMed]

132. Cohen, F.E.; Kelly, J.W. Therapeutic approaches to protein-misfolding diseases. *Nature* **2003**, *426*, 905–909. [CrossRef] [PubMed]

133. Haass, C.; Selkoe, D.J. Soluble protein oligomers in neurodegeneration: Lessons from the Alzheimer's amyloid β-peptide. *Nat. Rev. Mol. Cell Biol.* **2007**, *8*, 101–112. [CrossRef] [PubMed]

134. Cendrowska, U.; Silva, P.J.; Ait-Bouziad, N.; Müller, M.; Guven, Z.P.; Vieweg, S.; Chiki, A.; Radamaker, L.; Kumar, S.T.; Fändrich, M.; et al. Unraveling the complexity of amyloid polymorphism using gold nanoparticles and cryo-EM. *Proc. Natl. Acad. Sci. USA* **2020**, *117*, 6866–6874. [CrossRef] [PubMed]

135. Heegaard, P.M.H.; Boas, U.; Otzen, D.E. Dendrimer Effects on Peptide and Protein Fibrillation. *Macromol. Biosci.* **2007**, *7*, 1047–1059. [CrossRef] [PubMed]

136. Laumann, K.; Boas, U.; Larsen, H.M.; Heegaard, P.M.H.; Bergström, A.-L. Urea and Thiourea Modified Polypropyleneimine Dendrimers Clear Intracellular α-Synuclein Aggregates in a Human Cell Line. *Biomacromolecules* **2015**, *16*, 116–124. [CrossRef]

137. Holubová, M.; Štěpánek, P.; Hrubý, M. Polymer materials as promoters/inhibitors of amyloid fibril formation. *Colloid Polym. Sci.* **2020**. [CrossRef]

138. Lee, D.; Choe, Y.-J.; Lee, M.; Jeong, D.H.; Paik, S.R. Protein-Based SERS Technology Monitoring the Chemical Reactivity on an α-Synuclein-Mediated Two-Dimensional Array of Gold Nanoparticles. *Langmuir* **2011**, *27*, 12782–12787. [CrossRef]

139. Kim, C.-H.; Bhak, G.; Lee, J.; Sung, S.; Park, S.; Paik, S.R.; Yoon, M.-H. Controlled Charge Trapping and Retention in Large-Area Monodisperse Protein Metal-Nanoparticle Conjugates. *ACS Appl. Mater. Interfaces* **2016**, *8*, 11898–11903. [CrossRef]

140. Lee, D.; Hong, J.W.; Park, C.; Lee, H.; Lee, J.E.; Hyeon, T.; Paik, S.R. Ca^{2+}-Dependent Intracellular Drug Delivery System Developed with "Raspberry-Type" Particles-on-a-Particle Comprising Mesoporous Silica Core and α-Synuclein-Coated Gold Nanoparticles. *ACS Nano* **2014**, *8*, 8887–8895. [CrossRef]

141. Lee, D.; Choe, Y.-J.; Choi, Y.S.; Bhak, G.; Lee, J.; Paik, S.R. Photoconductivity of Pea-Pod-Type Chains of Gold Nanoparticles Encapsulated within Dielectric Amyloid Protein Nanofibrils of α-Synuclein. *Angew. Chem. Int. Ed.* **2011**, *50*, 1332–1337. [CrossRef] [PubMed]

142. Colby, R.; Hulleman, J.; Padalkar, S.; Rochet, J.C.; Stanciu, L.A. Biotemplated synthesis of metallic nanoparticle chains on an alpha-synuclein fiber scaffold. *J. Nanosci. Nanotechnol.* **2008**, *8*, 973–978. [CrossRef] [PubMed]

143. Kumar, J.; Eraña, H.; López-Martínez, E.; Claes, N.; Martín, V.F.; Solís, D.M.; Bals, S.; Cortajarena, A.L.; Castilla, J.; Liz-Marzán, L.M. Detection of amyloid fibrils in Parkinson's disease using plasmonic chirality. *Proc. Natl. Acad. Sci. USA* **2018**, *115*, 3225–3230. [CrossRef] [PubMed]

Publisher's Note: MDPI stays neutral with regard to jurisdictional claims in published maps and institutional affiliations.

Review

Protein Conformational Dynamics upon Association with the Surfaces of Lipid Membranes and Engineered Nanoparticles: Insights from Electron Paramagnetic Resonance Spectroscopy

Elka R. Georgieva

Department of Chemistry and Biochemistry, Texas Tech University, Lubbock, TX 79409, USA; elgeorgi@ttu.edu;
Tel.: +1-806-834-8166

Academic Editor: Michael Assfalg
Received: 26 October 2020; Accepted: 16 November 2020; Published: 18 November 2020

Abstract: Detailed study of conformational rearrangements and dynamics of proteins is central to our understanding of their physiological functions and the loss of function. This review outlines the applications of the electron paramagnetic resonance (EPR) technique to study the structural aspects of proteins transitioning from a solution environment to the states in which they are associated with the surfaces of biological membranes or engineered nanoobjects. In the former case these structural transitions generally underlie functional protein states. The latter case is mostly relevant to the application of protein immobilization in biotechnological industries, developing methods for protein purification, etc. Therefore, evaluating the stability of the protein functional state is particularly important. EPR spectroscopy in the form of continuous-wave EPR or pulse EPR distance measurements in conjunction with protein spin labeling provides highly versatile and sensitive tools to characterize the changes in protein local dynamics as well as large conformational rearrangements. The technique can be widely utilized in studies of both protein-membrane and engineered nanoobject-protein complexes.

Keywords: protein conformation; protein-surface association; lipid membranes; surface-immobilized protein; EPR spectroscopy

---

1. Introduction

Conformational dynamics is one of the key factors determining protein function [1,2]. Structural transitions, including large interdomain movements and local structure rearrangements, occur upon binding and processing of substrates [3,4], protein–protein interactions [5], and protein–nucleic acid interactions [6], as well as upon binding to the surface of biological membranes [7–10]. All these structure alterations are relevant to the physiological states of proteins and comprehensive understanding of them is central to developing approaches to monitor and control protein function. On the other hand, under certain not well understood conditions, some proteins adopt misfolded conformation in solutions or upon interaction with lipid membrane surfaces, leading to formation of large aggregates as those found in several neurodegenerative and other diseases [11,12]. Gaining deep understanding of protein conformation and stability is also very important in the cases where these proteins are immobilized on engineered surfaces. Protein immobilization through adsorption or covalent bonding is widely used in biotechnology, protein purification, drug delivery, etc. [13,14]. For such applications, it is important to characterize the efficiency and kinetics of protein immobilization, which significantly depends on the surface properties [15,16]. Furthermore, characterizing the conformational dynamics of proteins upon their immobilization on material surfaces serves to ensure that the protein functional states are preserved;

therefore, it is particularly important. Several techniques have been used to infer the structural dynamics of proteins upon association with biologically relevant surfaces, such as exhibited by lipid membranes, but also with engineered surfaces. Such techniques include, for example, FTIR spectroscopy [17], single-molecule fluorescence methods [18], and atomic force microscopy [19].

The emphasis of this review is on the application of electron paramagnetic resonance (EPR) spectroscopy in studies of protein structure and structural dynamics changes upon transition from the solution state to the surface-bound state. EPR in its continuous-wave (CW) and pulse spectroscopy modalities in combination with spin labeling provides the powerful biophysical techniques to study local dynamics and long-range conformational rearrangements of proteins [7,10,20–25]. Here, the EPR method with protein spin labeling is briefly described and is followed by the illustrations with relevant applications of this technique to study protein conformational transitions and structural stability upon binding to lipid membranes and engineered surfaces.

2. EPR Spectroscopy of Spin-Labeled Proteins

2.1. Spin Labeling of Protein Molecules with Nitroxides

Since EPR spectroscopy detects and studies paramagnetic centers which have unpaired electron spins, its applications to the structure and function of proteins having native paramagnetic centers is restricted to certain classes of metalloproteins [26,27] and radical enzymes [28–30]. Thus, protein molecules are predominantly diamagnetic and EPR silent. However, the range of proteins amenable to study by EPR spectroscopy is widely expanded by introducing paramagnetic reporter groups, so-called spin labels, which are covalently attached to desired positions in the protein molecules [20,31,32]. Therefore, EPR spectroscopy can be employed in studies of virtually any protein or protein complex, regardless of its size, environment (e.g., solution, lipid membrane), and flexibility (dynamics), given that it could be spin-labeled. The most universally used spin labels are nitroxides, which in most cases are introduced by covalent attachment of spin labeling reagent to the cysteine residue via disulfide or thioester bond formation. Most often they are mehanothyosulphonate- (MTS), iodoacetamido-, or maleimido-linked nitroxide tags [20,33] (Figure 1A–C). There is a large variety of less often used spin-bearing moieties as well. The cysteine residues used for spin labeling could be endogenous, but overwhelmingly they are introduced at desired positions through site-directed mutagenesis on the background of a cysteine-free protein (Figure 1D). Other types of spin labels based, for example, on trityl radicals [31,34,35], manganese ions [36], copper ions [37,38], gadolinium ions [34,36,39,40] have also been successful. In special cases spin labels are introduced by protein binding to spin-labeled substrate or inhibitor [41].

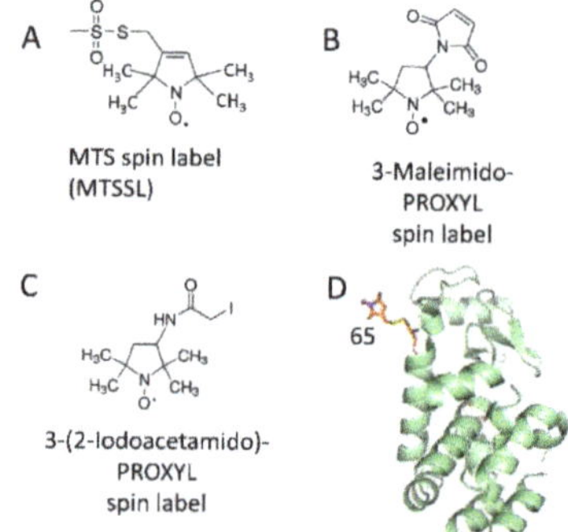

Figure 1. Spin labeling of proteins. (**A–C**) show nitroxide spin labels widely used for study of proteins by EPR spectroscopy. (**D**) emphasizes the residue 65 in T4 lysozyme spin-labeled with MTSSL. (PDB accession code 3LZM). Spin-label side chains were generated using the MMM program (Jeschke and Polyhach) [42].

2.2. CW and Pulse EPR Spectroscopy Applied to Spin-Labeled Proteins

Protein-bound nitroxide spin labels have been applied to study protein local dynamics and solvent accessibility at physiological temperatures by means of CW EPR spectroscopy [20,21,24,25,33,43–47]. Additionally, long-range structural rearrangements and local heterogeneities of biomolecules are best determined by using pulse EPR distance measurements between pairs of spin labels in a protein. The experiment is conducted as a rule at cryogenic temperatures but in few cases it was possible at physiological temperatures [7,9,10,20,22,25,28,33,45,46,48–51].

In nitroxide-labeled proteins lacking native paramagnetic centers, CW EPR records only the absorption spectrum of the nitroxide radical, usually in the form of the first derivative. The spectrum exhibits characteristic three lines, whose linewidths and amplitudes depend on the intrinsic spin label dynamics caused mainly by the rotational diffusion of the nitroxide moiety in the local potential and to some extent by the protein dynamics in its immediate vicinity as well as the protein global tumbling, segmental motion or chemical exchange [52–54]. In general, the frequencies of these motions observed by nitroxide multifrequency CW EPR fall into the range of rotational correlation time $(\tau_c) \cong 10^{-12}$–10^{-8} s. The anisotropy of hyperfine interaction (hfi) tensor and g-tensor, partly averaged by motions, produces typical asymmetric spectrum with the high-field side being broader. Visually the CW EPR spectrum is broader in the case of spin label with motion that is more restricted (Figure 2). As the diffusion rate decreases from fast to slow motion and is approaching rigid limit, the spectrum changes from three narrow lines with ~15 G splitting, to a more complex shape with broad linewidth until it ultimately approaches the rigid limit spectrum defined by just the magnetic tensors and, when present, by sample anisotropy. In the fast-motion regime, the rate of spin label dynamics is estimated from the central line width (ΔH) and the line intensity ratios. As the spectrum broadens, the outer line splitting in the spectrum (Figure 2) becomes useful for estimating the motional rate. As the motion slows down the spectrum becomes more complex [8,24,45] and may be significantly affected by the local ordering, e.g., microscopic-order-macroscopic-disorder (MOMD) cases [55]. The accurate determination of spin label ordering and dynamics necessitates resorting to sophisticated computational approaches that give rotational diffusion tensor and ordering parameters. This is an advantage to fast motions which yields only the rotational correlation time [55]. Therefore, changes in the spin label environment that produce steric effects, folding and unfolding, or binding to a much larger partner, such as a liposome or surface, all can be assessed through the rich variety of changes in the nitroxide CW EPR spectrum for selected spin-labeled positions. For example, if binding of a partner to a specific protein region decreases the rate of motion of this region, then one will record a CW EPR spectrum having broader lines compared to the EPR spectrum for protein without a binding partner [8,20,45]. Importantly, the motional rates corresponding to the slow-motion regime of spin-labels also depend on the strength of static magnetic field in a way that higher fields lead to more slow-motion spectra [56,57].

The degree of solvent exposure or penetration into the bilayer of certain membrane-bound protein regions can also be assessed by CW EPR spectroscopy. In this case, the residues in this region, with the exclusion of those highly conserved, are sequentially substituted for cysteines and labeled with a nitroxide. Then the CW EPR spectra of multiple spin-labeled cysteine variants are recorded in the presence of a paramagnetic relaxing agent (PRA), such as oxygen, nickel (II)–N,N1-ethylenediamine-diacetic acid (NiEDDA), or potassium ferricyanide ($K_3Fe(CN)_6$) [45,47,58,59]. The method relies on relaxation enhancement caused by the PRA changing the microwave (MW) power saturation properties of the nitroxide EPR spectrum. The experiment is referenced by one conducted in the absence of PRA. The intensity of the central peak in the nitroxide CW EPR spectrum is recorded as a function of the applied MW power, and the accessibility to the PRA is determined from the obtained power saturation curve [20,45,47,59]. By this method, one can assess the solvent exposure of spin-labeled residue at the membrane surface for certain protein conformations and determine whether this residue becomes more occluded or exposed upon conformational change. Furthermore, for proteins transitioning from soluble to membrane-bound state, one can establish which residues are in contact with the membrane surface or to what depth they may penetrate into the lipid bilayer [8,45,47].

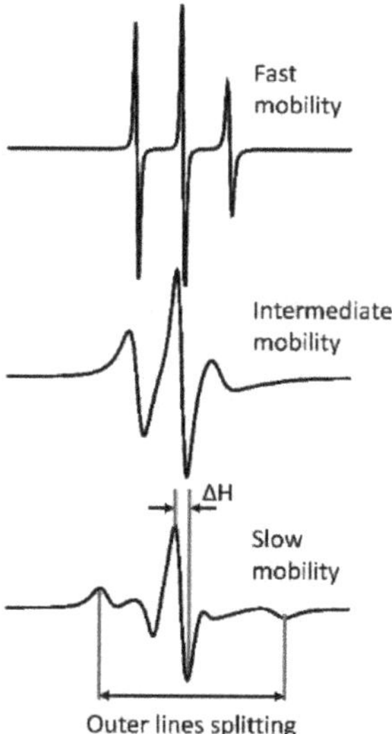

Figure 2. Continuous-wave (CW) electron paramagnetic resonance (EPR) spectra of the nitroxide spin label. The spectra correspond to different ranges of spin-label and protein motion. The coarse parameters, ΔH, and splitting of the outer hyperfine lines are shown in the bottom spectrum. Both parameters depend on the extent of spin label immobilization. The increase in the outer line splitting is due to the contribution from the A_{zz} hyperfine tensor component. Thus, less mobile spin labels lead to broader spectrum.

Pulse EPR spectroscopy with spin labeling also encompasses diverse applications in the realm of protein conformational rearrangements. The emphasis here is on the most widely used pulse EPR method, namely the four-pulse double electron–electron resonance (DEER) spectroscopy [22,23,48,60–62], which measures distances typically in the range of 2–7 nm and in special cases, e.g., in deuterated proteins, as long as 10 nm or somewhat more [7,63]. Other versions of pulse EPR, such as five-pulse DEER [64] and double quantum coherence [60,65,66] also have been very useful, particularly when increase in sensitivity or distance range was needed. The description of many other recently emerged technological developments in pulse distance EPR spectroscopy is beyond the scope of this review. All these pulse EPR methods measure the strength of the electron spin dipole–dipole interactions between coupled paramagnetic centers (predominantly covalently attached spin labels or intrinsic paramagnetic centers in proteins). The obtained direct information on the coupling strength ν_{dd} of these magnetic interactions (expressed in frequency units) also reports on the distance, r, between the paramagnetic spin-labels since $\nu_{dd} \propto 1/r^3$, (Figure 3). DEER spectroscopy has been used in numerous studies of protein conformation [51,63,67–73], many of which were carried out on membrane proteins [7,9,10,20,45,46,48,54,74–79]. The method is particularly suited for studies of conformational transitions taking place upon binding of protein molecules to lipid membrane surfaces, because very few methods can work well under the conditions of high protein heterogeneity or large flexibility, and the method is unconcerned about the typical large sizes of membrane mimetics. DEER measurements are commonly conducted at cryogenic temperatures. This was shown not to be a matter of concern. Moreover, recent developments of spin labels with longer relaxation times have made it possible to conduct experiments at room temperature [35,80].

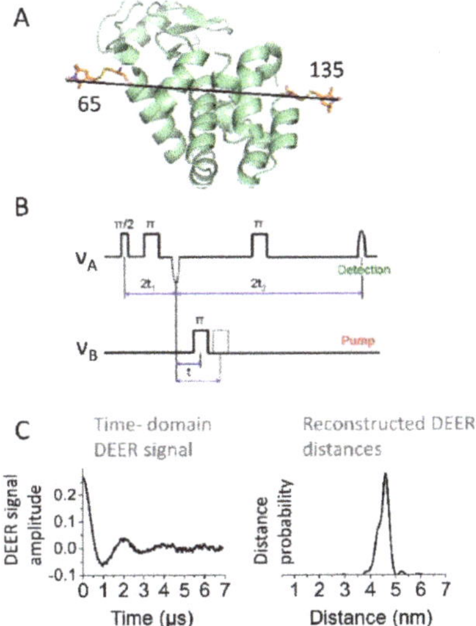

Figure 3. Double electron–electron resonance (DEER) distance measurements in doubly spin-labeled protein molecules. (**A**) The method is illustrated with T4 lysozyme (PDB accession code 3LZM) doubly spin-labeled at positions 65C and 135C. The spin label side-chains were generated with the program MMM. The measured inter-spin distance is indicated by the line. (**B**) The four-pulse microwave (MW) sequence that is used most often in the DEER experiment. The electron spins corresponding to different spin-labeled sites are frequency-selected for detection or pumping, respectively. The three MW pulses separated by fixed intervals select one of the spins to contribute to the detected spin-echo signal. The 4th (pump) pulse applied at a different frequency may select a coupled spin within the respective region of the spectrum. The pump pulse delay is advanced to give the spin-echo amplitude modulation envelope caused by dipolar coupling between the paired electron spins. (**C**) The recorded time-domain DEER signal (left) and the respective distances (right) between the MTSSL spins at residues 65C and 135C are shown. In general, a decaying oscillating time-domain signal is obtained from which the inter-spin distances and distance distributions are reconstructed.

3. EPR Studies Explore the Conformational Dynamics upon Interactions of Proteins with Biological and Engineered Surfaces

CW and pulse EPR spectroscopy techniques, applied to spin-labeled proteins, are well positioned to detect structural rearrangements that modify fast local and slower long-range dynamics of these complex molecules. Therefore, EPR methods have been extensively used to infer the information on physiologically relevant conformational changes of proteins caused, e.g., by transitioning from their soluble state to the membrane surface-bound state [7–10,33,45,46,81–84] but also aggregation and misfolding induced by the membrane surface [82,85,86]. EPR studies have also proved very useful in biotechnology developments for assessing protein stability when immobilized through absorption or covalent attachment to nanosurfaces and synthetic membranes [87–93]. This range of applications of EPR methods is highlighted with the examples provided below.

3.1. Conformational Dynamics Underlying Protein-Lipid Membrane Association

The high capacity of EPR methods to characterize protein dynamics and structural transitions in particularly from the soluble state to the membrane surface-associated state is highlighted here through the studies on human α-Synuclein, tau, and Snf7 proteins.

CW and pulse EPR spectroscopy both have been invaluable tools for studying the conformation changes accompanying the interaction of α-Synuclein with lipid membranes [7,8,10]. α-Synuclein, which is highly abundant in neurons, is a synaptic vesicle-associated protein with its key physiological function associated with the control over the trafficking of these vesicles and the release of neurotransmitters [94,95]. However, under not well-identified and understood conditions, this protein forms large insoluble aggregates found in the patients with severe neurodegenerations [95,96]. Due to the high importance of α-Synuclein in the physiology of the cell and neurodegeneration, a better understanding of the structural aspects underlying its function has been actively sought. However, the methods that can be applied are limited due to the high flexibility of the protein solution form and by the large size of membrane mimetics, further impeded by its structural heterogeneity in the case of the membrane-bound form. Circular dichroism (CD) spectroscopy and nuclear magnetic resonance (NMR) have established that α-Synuclein is unstructured and highly flexible in solutions [97,98]. However, its *N*-terminus and central regions adopt the fold with significant α-helical content upon binding to sodium dodecyl sulfate (SDS) micelles and other membrane mimetics [98–100]. EPR spectroscopy was particularly valuable as one of the main methods for overcoming the hurdles encountered by NMR and other biophysical and structural techniques. It provided detailed insights into the conformational rearrangements taking place when this protein binds to the surface of lipid membranes [7,8,10,101]. The studies have been performed on protein-spin label dynamics and on solvent accessibility of multiple singly spin-labeled residues by CW EPR and were advanced by carrying out long-range DEER distance measurements on several doubly spin-labeled α-Synuclein mutants. These CW EPR studies established that on the surface of liposomes, which closely represent the size and morphology of synaptic vesicles, the positively charged *N*-terminal and uncharged central region (also called the non-amyloid component, NAC) of α-Synuclein form an extended uninterrupted amphipathic α-helix. One side faces the solution environment, and the other side is associated with the liposome polar region [7,8,10,102]. This property was also determined for the α-Synuclein mutants found in patients with the familial form of Parkinson's disease [7]. EPR results have suggested that electrostatic and hydrophobic interactions stabilize the α-Synuclein-membrane complex [8,102]. The capacity of DEER spectroscopy to measure long distances up to nearly 9 nm in membrane-associated proteins, as well as the highly deuterated protein and lipids used in these studies made it possible to observe and characterize the extended helix nature of the binding *N*-terminal part of this protein [7,10]. In another DEER study on SDS and 1-palmitoyl−2-hydroxy-sn-glycero−3-phospho-(1′-rac-glycerol) (LPPG) micelle-bound α-Synuclein, it was found that the protein adopts the U-shape structure formed of two antiparallel amphipathic α-helices connected via an ordered linker [74], thereby confirming the NMR structure of the SDS-bound protein [100]. Next, EPR studies have revealed that the conformation of membrane-bound α-Synuclein can interconvert between the U-shaped antiparallel helices and the single extended helix. The structure depends on the protein-to-lipid (detergent) ratio. The high protein abundance promotes the U-shaped conformation and vice versa [7,75,103]. Thus EPR spectroscopy investigation on multiple spin-labeled single and double cysteine mutants of α-Synuclein, carried out on a variety of lipid membrane conditions, has culminated in a comprehensive characterization of the dynamic nature of this protein and highlighted its versatility conferred by adopting multiple membrane-bound conformations. This structural flexibility is likely much needed for the diverse physiological functions of α-Synuclein, but perhaps under yet unspecified conditions, the membrane association could trigger α-Synuclein misfolding and, possibly, aggregation [104]. It was further found by EPR that the *N*-terminus of membrane-associated α-Synuclein coordinates one equivalent of Cu^{2+} without any observed effect on protein helical structure, suggesting that Cu^{2+} uptake may be relevant to or be a part of α-Synuclein function [105]. Since Cu^{2+} is paramagnetic, the ion coordination was monitored by recording its CW EPR spectrum. The protein site of Cu^{2+} coordination was identified as Met1-Asp2 by using pulse EPR, namely electron spin echo envelope modulation, ESEEM, which reports on unpaired electron hyperfine interactions with the neighboring nuclei [106].

Conformational transitions upon binding to lipid membrane surfaces for another highly dynamic and physiologically important human protein tau were characterized using CW and pulse EPR spectroscopy [45]. A key physiological role of tau is to bind and stabilize microtubules, with the tau-microtubule association-dissociation equilibrium being regulated by tau phosphorylation [107,108].

However, when tau is hyperphosphorylated it transitions irreversibly into β-sheet structures that form large insoluble aggregates implicated in severe neurodegenerations [20,109]. In addition to its role in microtubule assembly and stabilization, this protein interacts with cellular membranes [110] with the physiological implications that are currently very little understood. The interaction with membranes in vitro was also linked to tau misfolding and aggregation [111], being in agreement with the detecting membrane-associated tau filaments in vivo [112]. Strikingly, tau associates with microtubules and lipid membranes using the same amino acid region known as microtubule binding domain (MBD). This domain consists of three or four imperfect repeats, depending on the protein isoform [113]. NMR and CD spectroscopies determined that this protein region, similarly to α-Synuclein, is highly unstructured in solution [114], but acquires helical conformation in the presence of membrane mimetics [115,116]. Yet, no clear understanding existed as to how the protein restructures upon binding to the membrane, as well as what its location is with respect to the membrane surface. These questions were answered in the EPR spectroscopy study [45]. Using CW EPR, multiple single spin-labeled mutants with cysteine residues spanning the entire tau MBD were scanned to determine which segments bind to liposomes and adopt a helical structure. To this end, EPR has determined the exposure of these spin-labeled cysteine residues to PRA in solvent/lipid accessibility measurements, as well as the alterations in the CW EPR spectrum lineshape underlying the spin label/cysteine residue dynamics. It was recognized that indeed tau associates with the membrane surface through its MBD and upon lipid binding this region transitions from highly dynamic to significantly more ordered conformation in which short α-helices are formed within each of the MBD repeats. Doubly spin-labeled mutants in the tau MBD were subjected to DEER spectroscopy to measure long-range distances between MBD repeats. The results, under the experimental conditions of this study, strongly suggest that the membrane-associated α-helices in the tau MBD are well separated, thus forming an extended structure, and these helices are connected via flexible linkers. Because no aggregation was observed, it is believed that the observed structures are physiologically relevant. The powerful combination of CW and pulse EPR methods provided a detailed characterization of the lipid-bound tau structure (Figure 4) [45], which provided the first comprehensive insights into how the large conformational dynamics in the tau MBD play a significant role in accommodating its interactions with its binding partners. Indeed, this is currently the most detailed structure of the tau MBD when bound to a physiological partner, and is deemed to be relevant to the structure of the tau protein in the microtubule-associated state.

In addition to intrinsically disordered proteins, which transition into significantly more defined partially folded conformations when peripherally bound to the membrane, EPR methods have been applied to elucidate the structural dynamics of more structured soluble proteins whose conformations drastically change to enable their association with membrane surfaces and possibly facilitating further insertion into the membrane [9,117,118]. Emphasized here is the study by DEER spectroscopy distance measurements and other biochemical and structural biology methods on the Snf7 protein from ESCRT-III (the endosomal sorting complexes required for transport) protein complex [9]. ESCRT III is a conformationally dynamic hetero-oligomer comprised of multiple units on the membrane. The complex plays essential role in membrane remodeling to support vital cellular processes [119–121]. Interestingly, each of the hetero-monomers is inactive in solution and possibly undergoes structural adjustments that enable protein-protein and protein-membrane surface interactions leading to formation of large spiral filaments. This conformational activation was spectacularly captured and characterized by DEER spectroscopy for the Snf7 protein [9]. In this study, DEER distance measurements were conducted on multiple single and doubly spin-labeled cysteine mutants of the Snf7 monomer. Experiments were performed for protein in solutions as well as in the presence of liposomes. The measured inter-spin distances within the Snf7 monomer provided direct evidence for protein restructuring. It was established that a certain region (helix 3) is highly dynamic in solutions, sampling between a conformation in close proximity to the core protein domain as well as more open conformation further away from the core domain. However, upon association with the liposome surface, helix 3 exclusively adopts the open conformation. This particular conclusion was based on the measured DEER distances and distance distributions for the pair of spin labels located in the helix 3 and in the core region of Snf7. Broad distance distributions were found in solutions, but they transformed to longer narrowly distributed distances for the lipid-bound protein. It was further found, based on the observed periodic DEER distances between single spin labels in the Snf7 monomer, that on the surface of the liposome, this protein forms filaments with ca. 30 Å periodicity. Thus, the study provided important insights into the structure dynamics underlying the mechanism of the activation of the Snf7 protein.

A

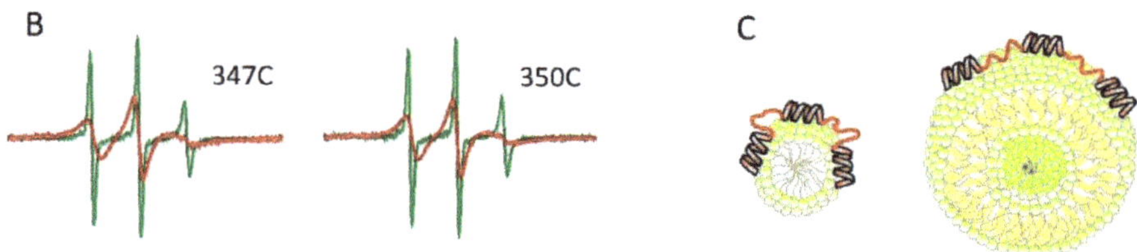

Environment	$r \pm \Delta r/2$						
	258C/273C helix1:linker1	258C/320C helix1:helix3	258C/352C helix1:helix4	273C/320C linker1:helix3	320C/335C helix3:linker2	320C/352C helix3:helix4	335C/352C Linker2:helix4
40 mM LPPG	2.4±0.7	3.5±1.1	5.0±1.3	3.4±0.8	2.5±0.7	3.3±1.2	2.8±1.1
POPC/POPS	2.6±0.9	6.0±1.7	N/A	3.8±1.1	2.5±0.7	5.5±1.5	3.7±1.3

B

C

Figure 4. The structure of membrane-bound tau protein as revealed by EPR spectroscopy is shown [45]. (**A**) Long inter-spin distances were measured by DEER spectroscopy for spin-labeled residues located in microtubule binding domain (MBD) repeats. The spin-labeled residues and their locations are indicated in the second row of the table. The lipid membrane mimetic (environment) in which the distances were obtained is indicated in the first column. The obtained distances (*r*) suggest that tau MBD has distinct structures when bound to LPPG micelles and 1-palmitoyl–2-oleoyl-glycero–3-phosphocholine/1-palmitoyl–2-oleoyl-sn-glycero–3-phospho-L-serine (POPC/POPS) liposomes. Longer distances in liposomes suggest more extended conformation. The distance distributions with relatively broad full widths at half maximum Δr indicated highly heterogeneous conformation of membrane-bound tau. (**B**) The binding of tau to lipid membranes was confirmed by CW EPR measurements. Nitroxide EPR lineshape broadening was observed for liposome-bound proteins (red spectra) compared to protein in solution (green spectra). The spin-labeled residues are indicated. (**C**) Model of tau-lipid membrane interaction based on the results from CW and DEER experiments was proposed—tau MBD structure when bound to micelles (left) and liposomes (right) is shown.

3.2. Conformational Dynamics and Structural Stability Underlying the Immobilization of Proteins on Engineered Surfaces

As pointed out above, the immobilization of functional proteins is of keen interest to technological developments in the biotechnological industries. To this end, proteins are immobilized through adsorption or covalent binding on the surfaces of engineered nanomaterial particles and synthetic membranes [58,89–93,122]. However, in some cases, there could be no compatibility between the surfaces (e.g., between the protein and pore sizes or due to charges or hydrophobicity), which can result in destabilization of the protein functional state leading to partial or the complete loss of function. The environmental factors, such as pH, could also affect protein binding and stability [123–125] Therefore, the need exists for robust methods of assessing the protein conformational changes associated with the immobilization on the surfaces of such materials. Provided here are the examples of how EPR spectroscopy helped in such assessments.

Berliner et al. applied the EPR method to study the conformations of trypsin active-site for protein in solution and when it is bound to porous glass [89]. In so doing, a spin-labeled compound 1-oxyl−2,2,6,6-tetramethyl−4-piperidinyl methylphosphonofluoridate, which act as trypsin inhibitor, was used. The mobility of this spin-labeled inhibitor when bound to the protein was studied using CW EPR spectroscopy. The CW EPR spectra recorded for protein in solution indicated the intermediate motional range of the spin label. A small additional spectral broadening was observed for the spin label bound to the immobilized enzyme due to the slower tumbling rate of the protein-glass bead complex. Importantly, identical EPR spectra were obtained for the immobilized protein interacted with the spin-labeled compound either before or after the immobilization. This suggests that the active site conformation in the immobilized trypsin is well preserved [89]. Certainly, this study spearheaded the applications of EPR spectroscopy to testing the structure of immobilized enzymes. Similar conclusions of having unaffected the active site of glutamate dehydrogenase immobilized on sepharose (a crosslinked, beaded-form of agarose) support were also drawn based on the results of the EPR study [93].

Clark and colleagues applied extensively CW EPR spectroscopy to study the effects of immobilization on the conformational dynamics of enzymes and enzyme active sites [90–92,126]. The technique was applied to horse liver alcohol dehydrogenase (LADH) to find the relationship between protein activity and its structure [126]. LADH, an oxoreductase homodimer, catalyzes the interconversions between aldehydes and alcohols and is active toward a broad spectrum of alcohols, aldehydes, and ketones. Therefore, the efforts have been made to understand how to optimize its application to industrial biotechnology. The effect of immobilization was studied [126] for two surfaces, namely CNBr-Sepharose (via covalent attachment) and Octyl-Sepharose (by adsorption). It was found that the enzyme-specific activity has decreased significantly upon immobilization, as compared with the enzyme in a solution; and the decrease was greater in the case of Octyl-Sepharose. Furthermore, the Octyl-Sepahrose-immobilized LADH had reduced thermal stability. CW EPR measurements on the spin-labeled enzyme helped to explain these results. In this regard, two approaches were utilized. Native zinc-coordinating cysteine residue at the catalytic site was labeled with an iodoacetamide-based nitroxide spin label, and the CW EPR spectra were recorded for the protein in free state in solution and when immobilized. Strikingly, two-component CW EPR spectra were recorded for both LADH in solution and immobilized LADH. Spectral decomposition was applied, which determined that the fractions of components corresponding to the slower and faster spin label motion were ca. 75% and 25%, respectively. Based on this, it was discovered that although LADH is a homodimer, with each protomer having one tightly coordinated zinc atom, the protomers in some LADH dimers adopt distinct conformations. Furthermore, the immobilization did not affect the populations of these states for both surfaces [126]. However, after the CW EPR spectra were decomposed, it become apparent that the spectral components for soluble and immobilized enzymes were not identical, with greater spectral broadening observed in the case of immobilized proteins. Thus, it was established that immobilization indeed affects the structure of LADH, but these structural changes were independent of the immobilization method. Based on the EPR data, the reduced thermal stability of the immobilized enzyme was interpreted as a result of conformational destabilization at the active site. For distinguishing between the two immobilization surfaces, a paramagnetic analog of 1,10-phenantroline (OPSL), which coordinates directly to the zinc atom at the LADH active site, was further utilized. It was found that the CW EPR spectrum of OPSL bound to the enzyme in solution was very close to those in the CNBr-Sepharose-immobilized enzyme (Figure 5). However, only 70% of OPSL was bound to the immobilized enzyme, which indicates that 30% of the active sites were inaccessible. Strikingly, a substantially different spectrum was recorded in the case of the Octyl-Sepharose-immobilized protein, providing evidence for an altered active site. These structural insights helped with the understanding of the structural reason underlying the reduced enzyme activity of immobilized LADH.

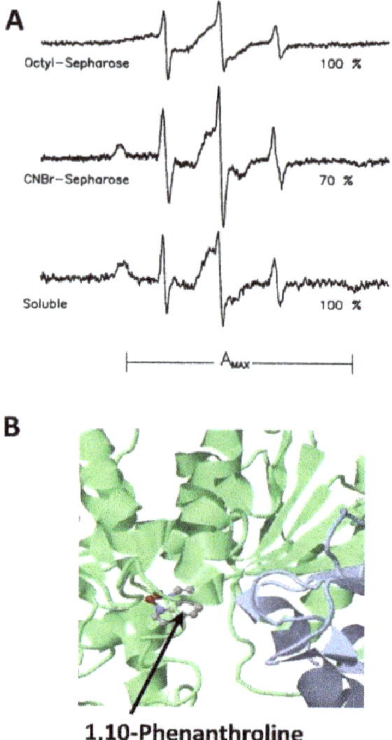

Figure 5. Conformations of liver alcohol dehydrogenase (LADH). (**A**) CW EPR spectra of spin-labeled OPSL bound to LADH in solution (bottom) and immobilized on two different carriers (mid and top). Tree sharp lines in each spectrum are from OPSL in solution with the rest of each spectrum is due to enzyme-bound OPSL. The numbers on the right represent the percentages of catalytic zinc ions and consequently of active sites available to OPSL. A_{MAX} corresponds to the outer spectral splitting, as indicated. (This figure from Ref. [126] is reproduced with permission from John Wiley; Sons) (**B**) A close-up view of 1,10-Phenanthroline bound to LADH at a putative binding site for OPSL, (PDB accession code 5VJ5).

CW EPR spectroscopy was also used to study the conformation of the papain enzyme, a cysteine protease, which was covalently immobilized on a fully hydrated porous polysulfone membrane [127] Native cysteine residue located in the cleft of the enzyme active site was labeled with an MTS-based nitroxide spin label and the CW EPR spectra were recorded on the spin-labeled protein in solution and when bound to the membrane. For the immobilized enzyme, these spectra consisted of two components with outer hyperfine splitting (which measures the spectral broadening and reports on the spin label/protein dynamics, Figure 2) of 60 G and 70 G, corresponding to an active site with faster and slower dynamics. The hyperfine splitting of 60 G was close to that of papain in solution (53 G). Nevertheless, both conformations in immobilized papain were different from those of free enzyme in solution, an indication that the immobilization indeed alters the structure of this protein. It was further found that the population of the less mobile component increased upon protein denaturation under varying pH, urea, or temperature conditions, as the larger hyperfine splitting of the 70 G component dominated the CW EPR spectrum. Thus, it was concluded that the second broader spectral component originated from protein, which was denatured due to binding to and immobilization onto the membrane. Next, for immobilized papain, experiments conducted to determine spin label accessibility to PRA $K_3Fe(CN)_6$ resulted in the paramagnetic broadening of only the more mobile component of the CW EPR spectrum. This indicates that this component corresponds to the functional enzyme, where the active site is solvent accessible. Therefore, the results from the CW EPR study were instrumental for understanding how and the extent to which the functional conformation is preserved when the papain enzyme is immobilized on an engineered membrane.

In more recent studies, pulse EPR distance measurements were conducted on sepharose-immobilized T4 lysozyme (T4L) with the goal of developing the applications of the method to studies at near-physiological temperatures [35]. To this end, double cysteine mutants of T4 lysozyme were labeled with a triarylmethyl (TAM)-based spin label. The spin label selection and protein immobilization provided the needed increase of the phase relaxation time and prevented from the rotational averaging of the anisotropic dipole-dipole interaction. This beneficial selection of experimental conditions made it possible to measure inter-spin distances in solutions at room temperature. This method could be further extended to study the long-range conformational stability of immobilized proteins relevant to technological developments in the enzyme industry, protein purification, etc.

4. Conclusions

EPR spectroscopy of spin-labeled proteins has demonstrated its effectiveness in characterizing the conformational states of proteins bound to physiologically relevant surfaces, such as those of lipid membranes. It also encompassed studies on protein binding to engineered surfaces, such as porous nanocarriers or synthetic membranes. In both cases, valuable information about functional protein conformations can be obtained. However, in contrast to rapidly expanding applications of this method in membrane protein structural and functional biology, its use in characterizing proteins for biotechnological or other industry and laboratory applications is lagging behind. These applications should be further developed to fully benefit from the high sensitivity of the CW EPR to changes in the local protein environment. This is possible through the measurements of local dynamics and solvent accessibility at the protein sites of interest. Furthermore, pulse EPR distance measurements offer an unparalleled capacity to provide deep insight into global protein stability and functionally relevant structural alterations.

Funding: This research received no external funding and was supported by startup funds (to E.R.G) from the Department of Chemistry and Biochemistry at TTU.

Acknowledgments: E.R.G. thanks Peter Borbat for critical reading of this manuscript and helpful suggestions. This work was supported by TTU Department of Chemistry and Biochemistry startup funds to E.R.G. The pulse EPR data in Figure 3C were obtained by E.R.G. at ACERT Cornell University, therefore she thanks for this opportunity.

Conflicts of Interest: The authors declare no conflict of interest.

References

1. Campitelli, P.; Modi, T.; Kumar, S.; Ozkan, S.B. The Role of Conformational Dynamics and Allostery in Modulating Protein Evolution. *Annu. Rev. Biophys.* **2020**, *49*, 267–288. [CrossRef] [PubMed]

2. Henzler-Wildman, K.; Kern, D. Dynamic personalities of proteins. *Nature* **2007**, *450*, 964–972. [CrossRef] [PubMed]

3. Seo, M.H.; Park, J.; Kim, E.; Hohng, S.; Kim, H.S. Protein conformational dynamics dictate the binding affinity for a ligand. *Nat. Commun.* **2014**, *5*, 3724. [CrossRef] [PubMed]

4. Kovermann, M.; Grundstrom, C.; Sauer-Eriksson, A.E.; Sauer, U.H.; Wolf-Watz, M. Structural basis for ligand binding to an enzyme by a conformational selection pathway. *Proc. Natl. Acad. Sci. USA* **2017**, *114*, 6298–6303. [CrossRef]

5. Chakrabarti, K.S.; Agafonov, R.V.; Pontiggia, F.; Otten, R.; Higgins, M.K.; Schertler, G.F.X.; Oprian, D.D.; Kern, D. Conformational Selection in a Protein-Protein Interaction Revealed by Dynamic Pathway Analysis. *Cell Rep.* **2016**, *14*, 32–42. [CrossRef] [PubMed]

6. Perez-Cano, L.; Eliahoo, E.; Lasker, K.; Wolfson, H.J.; Glaser, F.; Manor, H.; Bernado, P.; Fernandez-Recio, J. Conformational transitions in human translin enable nucleic acid binding. *Nucleic. Acids. Res.* **2013**, *41*, 9956–9966. [CrossRef]

7.   Georgieva, E.R.; Ramlall, T.F.; Borbat, P.P.; Freed, J.H.; Eliezer, D. The lipid-binding domain of wild type and mutant alpha-synuclein: Compactness and interconversion between the broken and extended helix forms. *J. Biol. Chem.* **2010**, *285*, 28261–28274. [CrossRef]

8.   Jao, C.C.; Der-Sarkissian, A.; Chen, J.; Langen, R. Structure of membrane-bound alpha-synuclein studied by site-directed spin labeling. *Proc. Natl. Acad. Sci. USA* **2004**, *101*, 8331–8336. [CrossRef]

9.   Tang, S.; Henne, W.M.; Borbat, P.P.; Buchkovich, N.J.; Freed, J.H.; Mao, Y.; Fromme, J.C.; Emr, S.D. Structural basis for activation, assembly and membrane binding of ESCRT-III Snf7 filaments. *eLife* **2015**, *4*, e12548. [CrossRef]

10. Georgieva, E.R.; Ramlall, T.F.; Borbat, P.P.; Freed, J.H.; Eliezer, D. Membrane-bound alpha-synuclein forms an extended helix: Long-distance pulsed ESR measurements using vesicles, bicelles, and rodlike micelles. *J. Am. Chem. Soc.* **2008**, *130*, 12856–12857. [CrossRef]

11. Dobson, C.M. Protein misfolding, evolution and disease. *Trends Biochem. Sci.* **1999**, *24*, 329–332. [CrossRef]

12. Soto, C.; Pritzkow, S. Protein misfolding, aggregation, and conformational strains in neurodegenerative diseases. *Nat. Neurosci.* **2018**, *21*, 1332–1340. [CrossRef] [PubMed]

13. Hlady, V.V.; Buijs, J. Protein adsorption on solid surfaces. *Curr. Opin. Biotechnol.* **1996**, *7*, 72–77. [CrossRef]

14. Gutierrez, R.; Valle, E.M.M.d.; Galan, M.A. Immobilized Metal-Ion Affinity Chromatography: Status and Trends. *Sep. Purif. Rev.* **2007**, *36*, 71–111. [CrossRef]

15. Mohamad, N.R.; Marzuki, N.H.; Buang, N.A.; Huyop, F.; Wahab, R.A. An overview of technologies for immobilization of enzymes and surface analysis techniques for immobilized enzymes. *Biotechnol. Biotechnol. Equip.* **2015**, *29*, 205–220. [CrossRef] [PubMed]

16. Jesionowski, T.; Zdarta, J.; Krajewska, B. Enzyme immobilization by adsorption: A review. *Adsorption* **2014**, *20*, 801–821. [CrossRef]

17. Shukla, R.J.; Singh, S.P. Structural and catalytic properties of immobilized alpha-amylase from Laceyella sacchari TSI-2. *Int. J. Biol. Macromol.* **2016**, *85*, 208–216. [CrossRef]

18. Kienle, D.F.; Falatach, R.M.; Kaar, J.L.; Schwartz, D.K. Correlating Structural and Functional Heterogeneity of Immobilized Enzymes. *ACS Nano* **2018**, *12*, 8091–8103. [CrossRef]

19. Marcuello, C.; de Miguel, R.; Gomez-Moreno, C.; Martinez-Julvez, M.; Lostao, A. An efficient method for enzyme immobilization evidenced by atomic force microscopy. *Protein Eng. Des. Sel.* **2012**, *25*, 715–723. [CrossRef]

20. Eschmann, N.A.; Georgieva, E.R.; Ganguly, P.; Borbat, P.P.; Rappaport, M.D.; Akdogan, Y.; Freed, J.H.; Shea, J.E.; Han, S. Signature of an aggregation-prone conformation of tau. *Sci. Rep.* **2017**, *7*, 44739. [CrossRef]

21. Hubbell, W.L.; Cafiso, D.S.; Altenbach, C. Identifying conformational changes with site-directed spin labeling. *Nat. Struct. Biol.* **2000**, *7*, 735–739. [CrossRef] [PubMed]

22. Jeschke, G. DEER distance measurements on proteins. *Annu. Rev. Phys. Chem.* **2012**, *63*, 419–446. [CrossRef] [PubMed]

23. Klare, J.P.; Steinhoff, H.J. Spin labeling EPR. *Photosynth. Res.* **2009**, *102*, 377–390. [CrossRef] [PubMed]

24. McHaourab, H.S.; Lietzow, M.A.; Hideg, K.; Hubbell, W.L. Motion of spin-labeled side chains in T4 lysozyme. Correlation with protein structure and dynamics. *Biochemistry* **1996**, *35*, 7692–7704. [CrossRef] [PubMed]

25. McHaourab, H.S.; Steed, P.R.; Kazmier, K. Toward the fourth dimension of membrane protein structure: Insight into dynamics from spin-labeling EPR spectroscopy. *Structure* **2011**, *19*, 1549–1561. [CrossRef]

26. More, C.; Belle, V.; Asso, M.; Fournel, A.; Roger, G.; Guigliarelli, B.; Bertrand, P. EPR spectroscopy: A powerful technique for the structural and functional investigation of metalloproteins. *Biospectroscopy* **1999**, *5*, S3–S18. [CrossRef]

27. Horitani, M.; Kusubayashi, K.; Oshima, K.; Yato, A.; Sugimoto, H.; Watanabe, K. X-ray Crystallography and Electron Paramagnetic Resonance Spectroscopy Reveal Active Site Rearrangement of Cold-Adapted Inorganic Pyrophosphatase. *Sci. Rep.* **2020**, *10*, 4368. [CrossRef]

28. Orlando, B.J.; Borbat, P.P.; Georgieva, E.R.; Freed, J.H.; Malkowski, M.G. Pulsed Dipolar Spectroscopy Reveals That Tyrosyl Radicals Are Generated in Both Monomers of the Cyclooxygenase-2 Dimer. *Biochemistry* **2015**, *54*, 7309–7312. [CrossRef]

29. Bennati, M.; Robblee, J.H.; Mugnaini, V.; Stubbe, J.; Freed, J.H.; Borbat, P. EPR distance measurements support a model for long-range radical initiation in *E. coli* ribonucleotide reductase. *J. Am. Chem. Soc.* **2005**, *127*, 15014–15015. [CrossRef]

30. Yee, E.F.; Diensthuber, R.P.; Vaidya, A.T.; Borbat, P.P.; Engelhard, C.; Freed, J.H.; Bittl, R.; Moglich, A.; Crane, B.R. Signal transduction in light-oxygen-voltage receptors lacking the adduct-forming cysteine residue. *Nat. Commun.* **2015**, *6*, 10079. [CrossRef]

31. Jassoy, J.J.; Heubach, C.A.; Hett, T.; Bernhard, F.; Haege, F.R.; Hagelueken, G.; Schiemann, O. Site Selective and Efficient Spin Labeling of Proteins with a Maleimide-Functionalized Trityl Radical for Pulsed Dipolar EPR Spectroscopy. *Molecules* **2019**, *24*, 2735. [CrossRef] [PubMed]

32. Hubbell, W.L.; McHaourab, H.S.; Altenbach, C.; Lietzow, M.A. Watching proteins move using site-directed spin labeling. *Structure* **1996**, *4*, 779–783. [CrossRef]

33. Klug, C.S.; Feix, J.B. Methods and applications of site-directed spin labeling EPR spectroscopy. *Methods Cell Biol.* **2008**, *84*, 617–658.

34. Giannoulis, A.; Yang, Y.; Gong, Y.J.; Tan, X.; Feintuch, A.; Carmieli, R.; Bahrenberg, T.; Liu, Y.; Su, X.C.; Goldfarb, D. DEER distance measurements on trityl/trityl and Gd(iii)/trityl labelled proteins. *Phys. Chem. Chem. Phys.* **2019**, *21*, 10217–10227. [CrossRef] [PubMed]

35. Yang, Z.; Liu, Y.; Borbat, P.; Zweier, J.L.; Freed, J.H.; Hubbell, W.L. Pulsed ESR dipolar spectroscopy for distance measurements in immobilized spin labeled proteins in liquid solution. *J. Am. Chem. Soc.* **2012**, *134*, 9950–9952. [CrossRef]

36. Yang, Y.; Gong, Y.J.; Litvinov, A.; Liu, H.K.; Yang, F.; Su, X.C.; Goldfarb, D. Generic tags for Mn(ii) and Gd(iii) spin labels for distance measurements in proteins. *Phys. Chem. Chem. Phys.* **2017**, *19*, 26944–26956. [CrossRef]

37. Merz, G.E.; Borbat, P.P.; Pratt, A.J.; Getzoff, E.D.; Freed, J.H.; Crane, B.R. Copper-based pulsed dipolar ESR spectroscopy as a probe of protein conformation linked to disease states. *Biophys. J.* **2014**, *107*, 1669–1674. [CrossRef]

38. Cunningham, T.F.; Putterman, M.R.; Desai, A.; Horne, W.S.; Saxena, S. The double-histidine Cu(2)(+)-binding motif: A highly rigid, site-specific spin probe for electron spin resonance distance measurements. *Angew. Chem. Int. Ed. Engl.* **2015**, *54*, 6330–6334. [CrossRef]

39. Yardeni, E.H.; Bahrenberg, T.; Stein, R.A.; Mishra, S.; Zomot, E.; Graham, B.; Tuck, K.L.; Huber, T.; Bibi, E.; McHaourab, H.S.; et al. Probing the solution structure of the E. coli multidrug transporter MdfA using DEER distance measurements with nitroxide and Gd(III) spin labels. *Sci. Rep.* **2019**, *9*, 12528. [CrossRef]

40. Galazzo, L.; Meier, G.; Timachi, M.H.; Hutter, C.A.J.; Seeger, M.A.; Bordignon, E. Spin-labeled nanobodies as protein conformational reporters for electron paramagnetic resonance in cellular membranes. *Proc. Natl. Acad. Sci. USA* **2020**, *117*, 2441–2448. [CrossRef]

41. Upadhyay, A.K.; Borbat, P.P.; Wang, J.; Freed, J.H.; Edmondson, D.E. Determination of the oligomeric states of human and rat monoamine oxidases in the outer mitochondrial membrane and octyl beta-D-glucopyranoside micelles using pulsed dipolar electron spin resonance spectroscopy. *Biochemistry* **2008**, *47*, 1554–1566. [CrossRef] [PubMed]

42. Polyhach, Y.; Bordignon, E.; Jeschke, G. Rotamer libraries of spin labelled cysteines for protein studies. *Phys. Chem. Chem. Phys.* **2011**, *13*, 2356–2366. [CrossRef] [PubMed]

43. Altenbach, C.; Marti, T.; Khorana, H.G.; Hubbell, W.L. Transmembrane protein structure: Spin labeling of bacteriorhodopsin mutants. *Science* **1990**, *248*, 1088–1092. [CrossRef] [PubMed]

44. Dimitrova, A.; Walko, M.; Hashemi Shabestari, M.; Kumar, P.; Huber, M.; Kocer, A. In situ, Reversible Gating of a Mechanosensitive Ion Channel through Protein-Lipid Interactions. *Front. Physiol.* **2016**, *7*, 409. [CrossRef] [PubMed]

45. Georgieva, E.R.; Xiao, S.; Borbat, P.P.; Freed, J.H.; Eliezer, D. Tau binds to lipid membrane surfaces via short amphipathic helices located in its microtubule-binding repeats. *Biophys. J.* **2014**, *107*, 1441–1452. [CrossRef] [PubMed]

46. Cooper, R.S.; Georgieva, E.R.; Borbat, P.P.; Freed, J.H.; Heldwein, E.E. Structural basis for membrane anchoring and fusion regulation of the herpes simplex virus fusogen gB. *Nat. Struct. Mol. Biol.* **2018**, *25*, 416–424. [CrossRef]

47. Altenbach, C.; Greenhalgh, D.A.; Khorana, H.G.; Hubbell, W.L. A collision gradient method to determine the immersion depth of nitroxides in lipid bilayers: Application to spin-labeled mutants of bacteriorhodopsin. *Proc. Natl. Acad. Sci. USA* **1994**, *91*, 1667–1671. [CrossRef]

48. Georgieva, E.R.; Borbat, P.P.; Ginter, C.; Freed, J.H.; Boudker, O. Conformational ensemble of the sodium-coupled aspartate transporter. *Nat. Struct. Mol. Biol.* **2013**, *20*, 215–221. [CrossRef]

49. Georgieva, E.R.; Borbat, P.P.; Norman, H.D.; Freed, J.H. Mechanism of influenza A M2 transmembrane domain assembly in lipid membranes. *Sci. Rep.* **2015**, *5*, 11757. [CrossRef]

50. Lai, A.L.; Clerico, E.M.; Blackburn, M.E.; Patel, N.A.; Robinson, C.V.; Borbat, P.P.; Freed, J.H.; Gierasch, L.M. Key features of an Hsp70 chaperone allosteric landscape revealed by ion-mobility native mass spectrometry and double electron-electron resonance. *J. Biol. Chem.* **2017**, *292*, 8773–8785. [CrossRef]

51. Puljung, M.C.; DeBerg, H.A.; Zagotta, W.N.; Stoll, S. Double electron-electron resonance reveals cAMP-induced conformational change in HCN channels. *Proc. Natl. Acad. Sci. USA* **2014**, *111*, 9816–9821. [CrossRef] [PubMed]

52. Freed, J.H. Theory of Slow Tumbling ESR Spectra for Nitroxides. In *Spin Labeling: Theory and Applications*; Berliner, L.J., Ed.; Elsevier: Amsterdam, The Netherlands, 1976.

53. Zhang, Z.; Fleissner, M.R.; Tipikin, D.S.; Liang, Z.; Moscicki, J.K.; Earle, K.A.; Hubbell, W.L.; Freed, J.H. Multifrequency electron spin resonance study of the dynamics of spin labeled T4 lysozyme. *J. Phys. Chem. B* **2010**, *114*, 5503–5521. [CrossRef] [PubMed]

54. Sahu, I.D.; Lorigan, G.A. Site-Directed Spin Labeling EPR for Studying Membrane Proteins. *BioMed Res. Int.* **2018**, *2018*, 3248289. [CrossRef] [PubMed]

55. Budil, D.E.; Lee, S.; Saxena, S.; Freed, J.H. Nonlinear-least-squares analysis of slow-motion EPR spectra in one and two dimentions using a modified Levenberg-Marquard algorithm. *J. Magn. Reson.* **1996**, *120*, 155–189. [CrossRef]

56. Borbat, P.P.; Costa-Filho, A.J.; Earle, K.A.; Moscicki, J.K.; Freed, J.H. Electron spin resonance in studies of membranes and proteins. *Science* **2001**, *291*, 266–269. [CrossRef]

57. Freed, J.H. New technologies in electron spin resonance. *Annu. Rev. Phys. Chem.* **2000**, *51*, 655–689. [CrossRef] [PubMed]

58. Butterfield, D.A.; Lee, J. Active site structure and stability of the thiol protease papain studied by electron paramagnetic resonance employing a methanethiosulfonate spin label. *Arch. Biochem. Biophys.* **1994**, *310*, 167–171. [CrossRef]

59. Altenbach, C.; Froncisz, W.; Hemker, R.; McHaourab, H.; Hubbell, W.L. Accessibility of nitroxide side chains: Absolute Heisenberg exchange rates from power saturation EPR. *Biophys. J.* **2005**, *89*, 2103–2112. [CrossRef]

60. Borbat, P.P.; Freed, J.H. Measuring distances by pulsed dipolar ESR spectroscopy: Spin-labeled histidine kinases. *Methods Enzymol.* **2007**, *423*, 52–116.

61. Borbat, P.P.; Freed, J.H. Pros and Cons of Pulse Dipolar ESR: DQC and DEER. *EPR Newsl.* **2007**, *17*, 21–33.

62. Pannier, M.; Veit, S.; Godt, A.; Jeschke, G.; Spiess, H.W. Dead-time free measurement of dipole-dipole interactions between electron spins. 2000. *J. Magn. Reson.* **2011**, *213*, 316–325. [CrossRef] [PubMed]

63. Schmidt, T.; Walti, M.A.; Baber, J.L.; Hustedt, E.J.; Clore, G.M. Long Distance Measurements up to 160 A in the GroEL Tetradecamer Using Q-Band DEER EPR Spectroscopy. *Angew. Chem. Int. Ed. Engl.* **2016**, *55*, 15905–15909. [CrossRef] [PubMed]

64. Borbat, P.P.; Georgieva, E.R.; Freed, J.H. Improved Sensitivity for Long-Distance Measurements in Biomolecules: Five-Pulse Double Electron-Electron Resonance. *J. Phys. Chem. Lett.* **2013**, *4*, 170–175. [CrossRef] [PubMed]

65. Borbat, P.P.; McHaourab, H.S.; Freed, J.H. Protein structure determination using long-distance constraints from double-quantum coherence ESR: Study of T4 lysozyme. *J. Am. Chem. Soc.* **2002**, *124*, 5304–5314. [CrossRef] [PubMed]

66. Ruthstein, S.; Ji, M.; Mehta, P.; Jen-Jacobson, L.; Saxena, S. Sensitive Cu^{2+}-Cu^{2+} distance measurements in a protein-DNA complex by double-quantum coherence ESR. *J. Phys. Chem. B* **2013**, *117*, 6227–6230. [CrossRef] [PubMed]

67. Reginsson, G.W.; Schiemann, O. Pulsed electron-electron double resonance: Beyond nanometre distance measurements on biomacromolecules. *Biochem. J.* **2011**, *434*, 353–363. [CrossRef]

68. Georgieva, E.R.; Roy, A.S.; Grigoryants, V.M.; Borbat, P.P.; Earle, K.A.; Scholes, C.P.; Freed, J.H. Effect of freezing conditions on distances and their distributions derived from Double Electron Electron Resonance (DEER): A study of doubly-spin-labeled T4 lysozyme. *J. Magn. Reson.* **2012**, *216*, 69–77. [CrossRef]

69. Huang, X.; de Vera, I.M.; Veloro, A.M.; Blackburn, M.E.; Kear, J.L.; Carter, J.D.; Rocca, J.R.; Simmerling, C.; Dunn, B.M.; Fanucci, G.E. Inhibitor-induced conformational shifts and ligand-exchange dynamics for HIV-1 protease measured by pulsed EPR and NMR spectroscopy. *J. Phys. Chem. B* **2012**, *116*, 14235–14244. [CrossRef]

70. Jeschke, G. The contribution of modern EPR to structural biology. *Emerg. Top. Life Sci.* **2018**, *2*, 9–18.

71. Schultz, K.M.; Fischer, M.A.; Noey, E.L.; Klug, C.S. Disruption of the E. coli LptC dimerization interface and characterization of lipopolysaccharide and LptA binding to monomeric LptC. *Protein Sci.* **2018**, *27*, 1407–1417. [CrossRef]

72. Selmke, B.; Borbat, P.P.; Nickolaus, C.; Varadarajan, R.; Freed, J.H.; Trommer, W.E. Open and Closed Form of Maltose Binding Protein in Its Native and Molten Globule State as Studied by Electron Paramagnetic Resonance Spectroscopy. *Biochemistry* **2018**, *57*, 5507–5512. [CrossRef] [PubMed]

73. Yin, D.M.; Hammler, D.; Peter, M.F.; Marx, A.; Schmitz, A.; Hagelueken, G. Inhibitor-Directed Spin Labelling—A High Precision and Minimally Invasive Technique to Study the Conformation of Proteins in Solution. *Chemistry* **2018**, *24*, 6665–6671. [CrossRef] [PubMed]

74. Borbat, P.; Ramlall, T.F.; Freed, J.H.; Eliezer, D. Inter-helix distances in lysophospholipid micelle-bound alpha-synuclein from pulsed ESR measurements. *J. Am. Chem. Soc.* **2006**, *128*, 10004–10005. [CrossRef] [PubMed]

75. Bortolus, M.; Tombolato, F.; Tessari, I.; Bisaglia, M.; Mammi, S.; Bubacco, L.; Ferrarini, A.; Maniero, A.L. Broken helix in vesicle and micelle-bound alpha-synuclein: Insights from site-directed spin labeling-EPR experiments and MD simulations. *J. Am. Chem. Soc.* **2008**, *130*, 6690–6691. [CrossRef]

76. Bordignon, E. Site-directed spin labeling of membrane proteins. *Top. Curr. Chem.* **2012**, *321*, 121–157.

77. Claxton, D.P.; Kazmier, K.; Mishra, S.; McHaourab, H.S. Navigating Membrane Protein Structure, Dynamics, and Energy Landscapes Using Spin Labeling and EPR Spectroscopy. *Methods Enzymol.* **2015**, *564*, 349–387.

78. Joseph, B.; Sikora, A.; Cafiso, D.S. Ligand Induced Conformational Changes of a Membrane Transporter in E. coli Cells Observed with DEER/PELDOR. *J. Am. Chem. Soc.* **2016**, *138*, 1844–1847. [CrossRef]

79. Kumar, P.; van Son, M.; Zheng, T.; Valdink, D.; Raap, J.; Kros, A.; Huber, M. Coiled-coil formation of the membrane-fusion K/E peptides viewed by electron paramagnetic resonance. *PLoS ONE* **2018**, *13*, e0191197. [CrossRef]

80. Meyer, V.; Swanson, M.A.; Clouston, L.J.; Boratynski, P.J.; Stein, R.A.; McHaourab, H.S.; Rajca, A.; Eaton, S.S.; Eaton, G.R. Room-temperature distance measurements of immobilized spin-labeled protein by DEER/PELDOR. *Biophys. J.* **2015**, *108*, 1213–1219. [CrossRef]

81. Snead, D.; Lai, A.L.; Wragg, R.T.; Parisotto, D.A.; Ramlall, T.F.; Dittman, J.S.; Freed, J.H.; Eliezer, D. Unique Structural Features of Membrane-Bound C-Terminal Domain Motifs Modulate Complexin Inhibitory Function. *Front. Mol. Neurosci.* **2017**, *10*, 154. [CrossRef]

82. Tao, M.; Pandey, N.K.; Barnes, R.; Han, S.; Langen, R. Structure of Membrane-Bound Huntingtin Exon 1 Reveals Membrane Interaction and Aggregation Mechanisms. *Structure* **2019**, *27*, 1570–1580.e4. [CrossRef] [PubMed]

83. Tao, M.; Isas, J.M.; Langen, R. Annexin B12 Trimer Formation is Governed by a Network of Protein-Protein and Protein-Lipid Interactions. *Sci. Rep.* **2020**, *10*, 5301. [CrossRef]

84. Bates, I.R.; Feix, J.B.; Boggs, J.M.; Harauz, G. An immunodominant epitope of myelin basic protein is an amphipathic alpha-helix. *J. Biol. Chem.* **2004**, *279*, 5757–5764. [CrossRef] [PubMed]

85. Rawat, A.; Langen, R.; Varkey, J. Membranes as modulators of amyloid protein misfolding and target of toxicity. *Biochim. Biophys. Acta Biomembr.* **2018**, *1860*, 1863–1875. [CrossRef]

86. Rawat, A.; Maity, B.K.; Chandra, B.; Maiti, S. Aggregation-induced conformation changes dictate islet amyloid polypeptide (IAPP) membrane affinity. *Biochim. Biophys. Acta Biomembr.* **2018**, *1860*, 1734–1740. [CrossRef]

87. Pan, Y.; Neupane, S.; Farmakes, J.; Bridges, M.; Froberg, J.; Rao, J.; Qian, S.Y.; Liu, G.; Choi, Y.; Yang, Z. Probing the structural basis and adsorption mechanism of an enzyme on nano-sized protein carriers. *Nanoscale* **2017**, *9*, 3512–3523. [CrossRef]

88. Jahnke, J.P.; Idso, M.N.; Hussain, S.; Junk, M.J.N.; Fisher, J.M.; Phan, D.D.; Han, S.; Chmelka, B.F. Functionally Active Membrane Proteins Incorporated in Mesostructured Silica Films. *J. Am. Chem. Soc.* **2018**, *140*, 3892–3906. [CrossRef]

89. Berliner, L.J.; Miller, S.T.; Uy, R.; Royer, G.P. An ESR study of the active-site conformations of free and immobilized trypsin. *Biochim. Biophys. Acta* **1973**, *315*, 195–199. [CrossRef]

90. Clark, D.S.; Bailey, J.E. Characterization of heterogeneous immobilized enzyme subpopulations using EPR spectroscopy. *Biotechnol. Bioeng.* **1984**, *26*, 231–238. [CrossRef]

91. Skerker, P.S.; Miller, R.R.; Millhauser, G.L.; Clark, D.S. Structural and functional responses of enzymes to immobilization: New insights from EPR spectroscopy. *Ann. N. Y. Acad. Sci.* **1987**, *501*, 80–84. [CrossRef]

92. Marg, G.A.; Millhauser, G.L.; Skerker, P.S.; Clark, D.S. Aplication of EPR method in studies of immobilized enzyme systems. *Ann. N. Y. Acad. Sci.* **1987**, *469*, 253–258. [CrossRef] [PubMed]

93. O'Connor, K.C.; Bailey, J.E. ESR investigations of free and immobilized glutamate dehydrogenase. *Biotechnol. Bioeng.* **1989**, *34*, 110–116. [CrossRef] [PubMed]

94. Snead, D.; Eliezer, D. Intrinsically disordered proteins in synaptic vesicle trafficking and release. *J. Biol. Chem.* **2019**, *294*, 3325–3342. [CrossRef] [PubMed]

95. Sulzer, D.; Edwards, R.H. The physiological role of alpha-synuclein and its relationship to Parkinson's Disease. *J. Neurochem.* **2019**, *150*, 475–486. [CrossRef]

96. Delenclos, M.; Burgess, J.D.; Lamprokostopoulou, A.; Outeiro, T.F.; Vekrellis, K.; McLean, P.J. Cellular models of alpha-synuclein toxicity and aggregation. *J. Neurochem.* **2019**, *150*, 566–576. [CrossRef]

97. Fauvet, B.; Mbefo, M.K.; Fares, M.B.; Desobry, C.; Michael, S.; Ardah, M.T.; Tsika, E.; Coune, P.; Prudent, M.; Lion, N.; et al. Alpha-Synuclein in central nervous system and from erythrocytes, mammalian cells, and Escherichia coli exists predominantly as disordered monomer. *J. Biol. Chem.* **2012**, *287*, 15345–15364. [CrossRef]

98. Eliezer, D.; Kutluay, E.; Bussell, R., Jr.; Browne, G. Conformational properties of alpha-synuclein in its free and lipid-associated states. *J. Mol. Biol.* **2001**, *307*, 1061–1073. [CrossRef]

99. Maiorano, J.N.; Davidson, W.S. The orientation of helix 4 in apolipoprotein A-I-containing reconstituted high density lipoproteins. *J. Biol. Chem.* **2000**, *275*, 17374–17380. [CrossRef]

100. Ulmer, T.S.; Bax, A.; Cole, N.B.; Nussbaum, R.L. Structure and dynamics of micelle-bound human alpha-synuclein. *J. Biol. Chem.* **2005**, *280*, 9595–9603. [CrossRef]

101. Weickert, S.; Cattani, J.; Drescher, M. Intrinsically disordered proteins (IDPs) studied by EPR and in-cell EPR. In *Electron Paramagnetic Resonance*; Chechik, V., Murphy, D.M., Eds.; Royal Society of Chemistry: London, UK, 2018; Volume 26, pp. 1–37.

102. Jao, C.C.; Hegde, B.G.; Chen, J.; Haworth, I.S.; Langen, R. Structure of membrane-bound alpha-synuclein from site-directed spin labeling and computational refinement. *Proc. Natl. Acad. Sci. USA* **2008**, *105*, 19666–19671. [CrossRef]

103. Drescher, M.; van Rooijen, B.D.; Veldhuis, G.; Subramaniam, V.; Huber, M. A stable lipid-induced aggregate of alpha-synuclein. *J. Am. Chem. Soc.* **2010**, *132*, 4080–4082. [CrossRef] [PubMed]

104. Alza, N.P.; Iglesias Gonzalez, P.A.; Conde, M.A.; Uranga, R.M.; Salvador, G.A. Lipids at the Crossroad of alpha-Synuclein Function and Dysfunction: Biological and Pathological Implications. *Front. Cell Neurosci.* **2019**, *13*, 175. [CrossRef] [PubMed]

105. Dudzik, C.G.; Walter, E.D.; Abrams, B.S.; Jurica, M.S.; Millhauser, G.L. Coordination of copper to the membrane-bound form of alpha-synuclein. *Biochemistry* **2013**, *52*, 53–60. [CrossRef] [PubMed]

106. Deligiannakis, Y.; Louloudi, M.; Hadjiliadis, N. Electron spin echo envelope modulation (ESEEM) spectroscopy as a tool to investigate the coordination environment of metal centers. *Coord. Chem. Rev.* **2000**, *204*, 1–112. [CrossRef]

107. Cleveland, D.W.; Hwo, S.Y.; Kirschner, M.W. Purification of tau, a microtubule-associated protein that induces assembly of microtubules from purified tubulin. *J. Mol. Biol.* **1977**, *116*, 207–225. [CrossRef]

108. Cleveland, D.W.; Hwo, S.Y.; Kirschner, M.W. Physical and chemical properties of purified tau factor and the role of tau in microtubule assembly. *J. Mol. Biol.* **1977**, *116*, 227–247. [CrossRef]

109. Honson, N.S.; Kuret, J. Tau aggregation and toxicity in tauopathic neurodegenerative diseases. *J. Alzheimers Dis.* **2008**, *14*, 417–422. [CrossRef]

110. Brandt, R.; Leger, J.; Lee, G. Interaction of tau with the neural plasma membrane mediated by tau's amino-terminal projection domain. *J. Cell Biol.* **1995**, *131*, 1327–1340. [CrossRef]

111. Elbaum-Garfinkle, S.; Ramlall, T.; Rhoades, E. The role of the lipid bilayer in tau aggregation. *Biophys. J.* **2010**, *98*, 2722–2730. [CrossRef]

112. Mena, R.; Edwards, P.C.; Harrington, C.R.; Mukaetova-Ladinska, E.B.; Wischik, C.M. Staging the pathological assembly of truncated tau protein into paired helical filaments in Alzheimer's disease. *Acta Neuropathol.* **1996**, *91*, 633–641. [CrossRef]

113. Goedert, M.; Spillantini, M.G.; Jakes, R.; Rutherford, D.; Crowther, R.A. Multiple isoforms of human microtubule-associated protein tau: Sequences and localization in neurofibrillary tangles of Alzheimer's disease. *Neuron* **1989**, *3*, 519–526. [CrossRef]

114. Eliezer, D.; Barre, P.; Kobaslija, M.; Chan, D.; Li, X.; Heend, L. Residual structure in the repeat domain of tau: Echoes of microtubule binding and paired helical filament formation. *Biochemistry* **2005**, *44*, 1026–1036. [CrossRef] [PubMed]

115. Barre, P.; Eliezer, D. Folding of the repeat domain of tau upon binding to lipid surfaces. *J. Mol. Biol.* **2006**, *362*, 312–326. [CrossRef] [PubMed]

116. Barre, P.; Eliezer, D. Structural transitions in tau k18 on micelle binding suggest a hierarchy in the efficacy of individual microtubule-binding repeats in filament nucleation. *Protein Sci.* **2013**, *22*, 1037–1048. [CrossRef] [PubMed]

117. Gallop, J.L.; Jao, C.C.; Kent, H.M.; Butler, P.J.; Evans, P.R.; Langen, R.; McMahon, H.T. Mechanism of endophilin N-BAR domain-mediated membrane curvature. *EMBO J.* **2006**, *25*, 2898–2910. [CrossRef] [PubMed]

118. Bleicken, S.; Classen, M.; Padmavathi, P.V.; Ishikawa, T.; Zeth, K.; Steinhoff, H.J.; Bordignon, E. Molecular details of Bax activation, oligomerization, and membrane insertion. *J. Biol. Chem.* **2010**, *285*, 6636–6647. [CrossRef]

119. Henne, W.M.; Buchkovich, N.J.; Zhao, Y.; Emr, S.D. The endosomal sorting complex ESCRT-II mediates the assembly and architecture of ESCRT-III helices. *Cell* **2012**, *151*, 356–371. [CrossRef]

120. Henne, W.M.; Buchkovich, N.J.; Emr, S.D. The ESCRT pathway. *Dev. Cell* **2011**, *21*, 77–91. [CrossRef]

121. Hurley, J.H.; Hanson, P.I. Membrane budding and scission by the ESCRT machinery: It's all in the neck. *Nat. Rev. Mol. Cell Biol.* **2010**, *11*, 556–566. [CrossRef]

122. Bhardwaj, A.; Lee, J.B.; Glauner, K.; Ganapathi, S.; Bhattacharyya, D.; Butterfield, D.A. Biofunctional membranes: An EPR study of active site structure and stability of papain non-covalently immobilized on the surface of modified poly(ether) sulfone membranes through the avidin-biotin linkage. *J. Membr. Sci.* **1996**, *119*, 241–252. [CrossRef]

123. Ballet, T.; Boulange, L.; Brechet, Y.; Bruckert, F.; Weidenhaupt, M. Protein conformational changes induced by adsorption onto material surfaces: An important issue for biomedical applications of material science. *Bull. Pol. Acad. Sci.* **2010**, *58*, 303–315. [CrossRef]

124. Dee, K.C.; Puleo, D.A.; Bizios, R. Protein-Surface Interactions. In *An Introduction to Tissue-Biomaterial Interactions: Tissue-Biomaterial*; Dee, K.C., Puleo, D.A., Bizios, R., Eds.; Wiley-Liss, Inc.: New York, NY, USA, 2002; pp. 37–52.

125. Guzik, U.; Hupert-Kocurek, K.; Wojcieszynska, D. Immobilization as a strategy for improving enzyme properties-application to oxidoreductases. *Molecules* **2014**, *19*, 8995–9018. [CrossRef] [PubMed]

126. Clark, D.S.; Skerker, P.S.; Fernandez, E.J.; Jagoda, R.B. Spectroscopic studies of structure-function relationships in free and immobilized alcohol dehydrogenase. *Ann. N. Y. Acad. Sci.* **1987**, *506*, 117–128. [CrossRef] [PubMed]

127. Butterfield, D.A.; Lee, J.; Ganapathi, S.; Bhattacharyya, D. Biofunctional membranes 4. Active-site structure and stability of an immobilized enzyme, papain, on modified polysulfone membranes studied by electron-paramagnetic-resonance and kinetics. *J. Membr. Sci.* **1994**, *91*, 47–64. [CrossRef]

Sample Availability: Samples of the compounds are not available from the authors.

Publisher's Note: MDPI stays neutral with regard to jurisdictional claims in published maps and institutional affiliations.

 *molecules*

Article

Insights into a Protein-Nanoparticle System by Paramagnetic Perturbation NMR Spectroscopy

Yamanappa Hunashal [1,2], Cristina Cantarutti [3], Sofia Giorgetti [4], Loredana Marchese [4], Federico Fogolari [5,6] and Gennaro Esposito [1,6,*]

1   Science Division, New York University Abu Dhabi, 129188 Abu Dhabi, UAE; yh45@nyu.edu
2   Dipartimento di Area Medica, Università di Udine, P.le Kolbe, 4, 33100 Udine, Italy
3   Institute de Chimie, UMR CNRS 7272, Université Côte d'Azur, Université de Nice Sophia Antipolis, Parc Valrose, 06108 Nice, CEDEX 2, France; cristina.cantarutti@gmail.com
4   Dipartimento Medicina Molecolare, Università di Pavia, Via Taramelli 3, 27100 Pavia, Italy; sofia.giorgetti@unipv.it (S.G.); loredana.marchese@unipv.it (L.M.)
5   Dipartimento di Matematica, Informatica e Fisica, Università di Udine, Viale delle Scienze, 33100 Udine, Italy; federico.fogolari@uniud.it
6   Instituto Nazionale Biostrutture e Biosistemi, Viale Medaglie d'Oro 305, 00136 Roma, Italy
*   Correspondence: rino.esposito@nyu.edu; Tel./Fax: +97-126-2845-86

Academic Editors: Michael Assfalg, Roberto Fattorusso and Mark von Itzstein
Received: 6 July 2020; Accepted: 22 October 2020; Published: 7 November 2020

Abstract: Background: The interaction between proteins and nanoparticles is a very relevant subject because of the potential applications in medicine and material science in general. Further interest derives from the amyloidogenic character of the considered protein, β2-microglobulin (β2m), which may be regarded as a paradigmatic system for possible therapeutic strategies. Previous evidence showed in fact that gold nanoparticles (AuNPs) are able to inhibit β2m fibril formation in vitro. Methods: NMR (Nuclear Magnetic Resonance) and ESR (Electron Spin Resonance) spectroscopy are employed to characterize the paramagnetic perturbation of the extrinsic nitroxide probe Tempol on β2m in the absence and presence of AuNPs to determine the surface accessibility properties and the occurrence of chemical or conformational exchange, based on measurements conducted under magnetization equilibrium and non-equilibrium conditions. Results: The nitroxide perturbation analysis successfully identifies the protein regions where protein-protein or protein-AuNPs interactions hinder accessibility or/and establish exchange contacts. These information give interesting clues to recognize the fibrillation interface of β2m and hypothesize a mechanism for AuNPs fibrillogenesis inhibition. Conclusions: The presented approach can be advantageously applied to the characterization of the interface in protein-protein and protein-nanoparticles interactions.

Keywords: protein-nanoparticle interactions; protein NMR; amyloidogenic proteins; nitroxide paramagnetic perturbation; spin label extrinsic probes; Tempol; β2-microglobulin

---

1. Introduction

The interaction of proteins with nanoparticle (NP) systems is a very challenging issue that has many implications in physical chemistry as well as in biomedical and biochemical applications [1]. Depending on the NP charge, size, shape, and chemical functions on the surface, proteins may be adsorbed onto that surface, concentrated in a layer named corona, or experience labile interactions and exchange with bulk solution. There are several experimental strategies that can be employed to assess the interaction between NP and proteins, ranging from direct inspection by microscopy or suitable spectroscopic techniques, e.g., UV, fluorescence, surface-enhanced Raman spectroscopy, etc., to the indirect inference based on the assay of the protein function through the related biological or cellular response. More detailed information can be obtained also by other techniques such as NMR, although the application viability is restricted to those systems where the protein exchange can be exploited to gain information on the transiently bound states that are unobservable by NMR because of the large sizes. However, magnetic resonance can be employed over a wider scale range if electron resonance is considered, provided the systems under consideration respond to free paramagnetic species or can host suitable paramagnetic probes on the proteins or the NPs. For NMR applications, the most convenient NP size window to modulate the protein interaction is the medium-size range, i.e., 5–20 nm, where the effects of the NP shape, charge, and surface chemistry can be tuned for the scopes of interest.

Over the last few years, we have carried out systematic investigations on the interaction of amyloidogenic protein models and citrate-coated or alkanethiolate-coated gold NPs (AuNPs) with diameters of 3.6, 5, and 7.5 nm [2–7]. We worked in particular on β2-microglobulin (β2m) and variants thereof that represent a paradigmatic example of amyloidogenic protein misfolding [8,9]. β2m naturally occurs in class I major histocompatibility complex on the surface of antigen presenting cells, in conjunction with a larger domain. Due to renal failure and consequent high concentration from impaired clearance [8], or because of a mutation [9], a pathologic fibrillar aggregation of β2m takes place, leading to amyloid deposition in patients undergoing long-term hemodialysis or aged individuals with genetically inherited mutation.

NP interaction studies were conducted on wild-type β2m [10–12], the naturally occurring amyloidogenic mutant D76N β2m [9,13] and ΔN6 β2m, which is a variant devoid of the first six residues that is found only in natural fibrils of the wild-type species [14]. Contrary to the expectations based on earlier results [15], the protein solutions with AuNPs were stable for several months, and no evidence of increased aggregation or partial unfolding was observed. The occurrence of uneven patterns of signal attenuation was indicative of a preferential interface of fast exchange with AuNPs [2–4]. With D76N β2m, the most amyloidogenic variant of β2m that fibrillates by agitation at neutral pH, the presence of citrate-stabilized AuNPs inhibited fibrillogenesis by interfering with the early aggregation steps of the protein that are crucial for the protofibril nucleus formation, as inferred from NMR, QCMD (Quartz Crystal Microbalance with Dissipation monitoring), and MD (Molecular Dynamics) [3,4,6].

Recently, we have revived the use of water-soluble nitroxides such as Tempol to explore the exchange dynamics of β2m [16]. Tempol and similar stable free radicals had been formerly employed as extrinsic probes for identifying the protein exposed locations, based on the paramagnetic perturbation of the NMR signals induced by the unpaired electron of the radical over accessible molecular surfaces [17–20]. The same paramagnetic perturbation measured under non-equilibrium conditions of the NMR magnetization determines an attenuation pattern that differs from the corresponding profile obtained under equilibrium conditions, i.e., with fully relaxed NMR magnetization. While in the latter conditions, the extent of NMR attenuation reflects the proximity to the unpaired electron and therefore the accessibility of or the distance from the molecular surface, the attenuation retrieved under non-equilibrium conditions of magnetization recovery can map also the locations of hindered accessibility or μs-to-ms exchange events, by identifying slower or faster relaxing nuclei, respectively, with respect to the average relaxation rate enhancement brought about by the nitroxide probe [16].

Here, we show the application of this novel use of Tempol attenuation to gain insights into the interactions that wild-type β2m establishes with citrate-coated AuNPs. The analysis of the ternary system protein/AuNP/spin-label probe is conducted with respect to all the NMR- (Nuclear Magnetic Resonance) and ESR- (Electron Spin Resonance) accessible controls involving only two components of the system, which are based also on the previously reported evidence [2–7,16].

2. Results

Using extrinsic spin labels such as nitroxides to extract structural information requires testing the reliability of their non-specific probe behavior [17–20]. ESR spectra of Tempol in the absence and presence of β2m had previously shown that only statistical encounters occur between the free radical and the protein, as inferred from the invariance of linewidths and amplitudes of the superimposed spectra [16]. For the ternary system protein + AuNPs + Tempol, the ESR trace superposition for the three controls (Tempol, Tempol + AuNPs, and Tempol + β2m) and the ternary system shows substantial coincidence with some small amplitude deviations (Figure S1). The degree of meaningfulness of these deviations was assessed by calculating the rotational correlation time (τ_c) of the nitroxide in the different tested conditions, according to Equations (4) and (5) (see Section 4). Table 1 lists the corresponding values. Under any tested condition, the tempol τ_c value remains around an average of 31.6 ps (the standard deviation is 1.3 ps). This indicates that, within the experimental error, no detectable effect arises from β2m, or AuNPs, or both on the tumbling rate of the free radical. On the other hand, that average τ_c value is consistent with those reported for 2 mM Tempo (91.9 ps) and Tempone (14.9 ps) in water at 300 K [21]. Therefore, the occurrence of Tempol interactions other than the statistical collision in the binary and ternary systems here considered should be ruled out.

Table 1. Rotational correlation time ($\tau_c/10^{-11}$ s) of Tempol at the indicated concentrations and different solution compositions obtained from ESR measurements at 298 K.

Composition	Tempol	Tempol + AuNPs	Tempol + β2m	Tempol + AuNPs + β2m
[Tempol]		[AuNP] = 60 nM	[β2m] = 8 μM	[AuNP] = 60 nM; [β2m] = 8 μM
1.6 mM	3.3 ± 0.2	3.1 ± 0.2	3.4 ± 0.2	3.3 ± 0.2
0.8 mM	3.2 ± 0.2	3.0 ± 0.3	3.2 ± 0.3	2.9 ± 0.2
0.4 mM	3.1 ± 0.3	3.2 ± 0.3	3.1 ± 0.3	3.1 ± 0.4

Figure 1 shows the pattern of the normalized attenuation (A_N) values observed with 8 μM β2m and 0.8 mM Tempol with respect to the same protein solution without the nitroxide. The numerical values are listed in Table S1, along with the corresponding errors. The graph of Figure 1 depicts in red the backbone amide signal A_N values extracted from data collected under magnetization equilibrium conditions (A_N[eq]), and in blue the analogous A_N values extracted from data collected under magnetization off-equilibrium conditions (A_N[off-eq]), which were respectively obtained from pairs of ^{15}N-^{1}H HSQC spectra acquired with relaxation delays of 5 s and 0.5 s. According to our previous interpretation [16–18], A_N[eq] values larger or smaller than unity indicate amide signals attenuated above or below the average attenuation, respectively, and therefore, they identify molecular locations more or less accessible to the nitroxide probe, depending on the specific surface exposure. Instead, the interpretation of the A_N[off-eq] values is related to their relationship with the corresponding A_N[eq] figures [16]. In particular, the pattern A_N[off-eq] > A_N[eq] identifies amide positions with either locally hindered accessibility on the molecular surface or true structurally buried positions. As such, this pattern, which was named the Type I deviation of A_N[off-eq], is typically, though not exclusively, associated with A_N[eq] < 1, i.e., locations with accessibility lower than average [16]. As a matter of fact, when Type I deviation occurs at exposed locations, the corresponding A_N[eq] value is only slightly larger than unity. On the other hand, the pattern A_N[off-eq] < A_N[eq], named Type II deviation of A_N[off-eq], identifies those amide positions whose recovery is faster than the average off-equilibrium signal recovery, thereby proving less attenuated than that average. In the absence of specific interactions of the spin probe with the protein and/or structural transitions of the latter induced by the former, as verifiable by the invariance of the amide signal chemical shifts (Figure S2), Type II deviation of A_N[off-eq] can be associated to local chemical or conformational exchange occurring on a ms-to-μs time scale that introduces additional relaxation increments affecting both T_{1p} and T_{2p}, i.e., the paramagnetic contribution to longitudinal and transverse relaxation times [16].

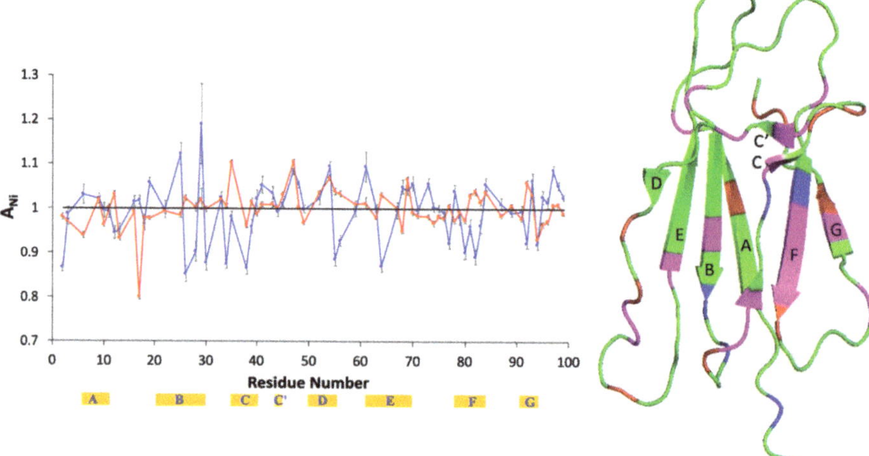

Figure 1. Overlay of the A_N values obtained from ^{1}H-^{15}N HSQC spectra of 8 μM β2m in the presence of 0.8 mM Tempol, with a relaxation delay of 0.5 s (**blue**) or 5 s (**red**). The β-strand location and naming along the sequence is reported with yellow strips. The cartoon on the right highlights the positions of the accessible amides (red), i.e., exhibiting A_N[eq] > 1, and the amides with Type II deviation of A_N[off-eq] (**blue**), i.e., displaying A_N[off-eq] < A_N[eq]. The magenta color denotes sites where both A_N[eq] > 1 and Type II deviation occur simultaneously. Here and elsewhere, the reproduced structure is the NMR solution structure of β2m [10] (Protein Data Bank or PDB code 1JNJ). The secondary structure elements of β2m are indicated according to the crystallographic naming scheme (PDB code 3HLA). Structures are always drawn with PyMOL (Schrödinger, Inc., version 2.3.5, New York, NY, USA).

The pattern of Figure 1 is different with respect to the corresponding one previously observed in diluted conditions, precisely at β2m concentration of 50 μM probed with 250 μM Tempol [16]. Apart from the larger error affecting the previous data, there are two important differences to point out. First, the former tempol/protein ratio was 5:1, whereas here, we consider a ratio of 100:1. These ratios and the absolute concentrations affect directly the collision probability [22]. We reasoned that a high tempol/protein ratio is required to balance a low absolute concentration of β2m and measure the paramagnetic perturbation. This is confirmed by the slight increase above the unity of the average relative intensity, RI_{av}, (for RI definition see Section 4) under off-equilibrium conditions [16] (Figure S3).

Second, at 8 μM concentration, the extent of β2m dimerization and higher oligomerization should be further reduced compared to 50 μM [14,23,24]. Hence, it could be possible to observe features related to the protein association interface. Table 2 lists the details of the pattern observed in Figure 1 compared to the earlier results at 50 μM [16]. The most relevant differences concern the higher exposure in the 8 μM solution of strands C, D, F, and G and the intensification in local conformational or chemical exchange at strands C and F, with a simultaneous loss of accessibility at strand A and loop AB. In addition to the relevance for the involvement in the association interface, these features are also important to delineate a starting point and thus appreciate the interactions and structural effects that the presence of AuNPs may induce.

Table 2. Paramagnetic perturbation induced by 0.8 mM Tempol on the amide NMR signals of 8 μM β2m. Equilibrium (column 2) and off-equilibrium (columns 3 and 4) data are compared to the corresponding data obtained at 50 μM β2m and 5:1 Tempol:protein ratio [16] and reported below in italic fonts.

Structure Region	$A_N > 1$	Type I A_N [off-eq] > A_N [eq]	Type II A_N [off-eq] < A_N [eq]
N-term, strand A	V9 *Q2, I7*	K6, Y10	Q2 *Q2, I7*
loop AB	*R12, K19*	H13, E16, N17, K19 *N17*	R12 *R12, H13*
strand B	Y26	C25, G29 *S28*	Y26, S28, F30 *N21, F22, C25, F30*
loop BC	S33, D34 *D34*	S33	D34 *D34*
strand C	I35, L39, K41 *I35*	L40, K41	I35, D38, L39 *I35*
turn CC', strand C', loop C'D	G43, R45, E47 *G43, E47, K48, V49*	G43, K48, V49 *R45, K48*	R45, E47 *N42*
strand D	S52, L54, S55, F56 *F56*	L54 *E50, L54*	S55, F56 *S52, F56*
loop DE	D59	S61 *D59*	*S61*
strand E	L64, E69 *E69*	Y63, T68, F70 *E69, F70*	L64, E69 *L64*
loop EF	E77 *K75*	T73, E74, K75, D76 *T71, E74*	E77
strand F	R81, V82, N83, H84 *N83, H84*	T78 *T78*	C80, R81, V82, N83 *A79, H84*
loop FG	Q89 *V85, L87*	L87	Q89 *V85*
strand G, C-term	I92, V93, R97, D98 *K94*	V93, W95, D96, R97, D98, M99 *V93, D96, D98, M99*	I92 *I92, K94*

The same paramagnetic perturbation analysis as done with isolated β2m can be performed for the system protein + AuNPs, because a statistical collision model can still be adopted, according to the ESR-based determinations of the Tempol τ_c values under different experimental conditions. Moreover, AuNPs are known to essentially preserve the chemical shifts, and therefore the structure, of β2m and D76N β2m [2,4,6], although at concentrations and NP/protein ratios as low as 4–8 μM and 1/100–1/200, small chemical shift deviations have been detected for both variants [2,6]. Most of these deviations were observed with synthetic AuNPs with an average diameter of 7.5 ± 1 nm, and therefore, some difference can be expected upon decreasing the NP diameter to 5 nm. With the commercial AuNPs here employed, minor chemical shift perturbations [25] are measured at Q2, N17, S33, D38, and S61 NHs of β2m, which in two cases (N17, D38) decrease below the resolution significance in the presence of Tempol (Figure S2). A possible explanation for those chemical shift perturbations may be related to NP-induced alterations of the intra-residue interaction between the side-chain polar group and the backbone amides. From the results reported in Figure 2 and Table 3, it can be seen that N17 and D38 become accessible to Tempol in the presence of AuNPs. Therefore, the reduction of their chemical shift perturbation could be related to the interaction with Tempol that, albeit non-specific, competes with the intra-residue interaction. On the other hand, the conserved chemical shift deviations of Q2, S33, and S61 after Tempol addition match with a hindered accessibility in the presence of AuNPs (Figure 2, Table 3). Therefore, all of the observed chemical shift perturbations can be attributed exclusively to the protein interaction with the NPs. However, given the substantial invariance of the fingerprint pattern in the ^{15}N-^{1}H HSQC spectra and the limited amounts of the frequency changes, all the mentioned deviations do not impair the assumptions of protein structure conservation and statistical character for the nitroxide probing. However, all the previous evidence indicates that AuNPs unevenly affect the intensity of the ^{15}N-^{1}H HSQC peaks of β2m and variants thereof [2–7]. Although this effect is precious to interpret the molecular details of the protein interaction with nanoparticles, it may prove detrimental when evaluating the paramagnetic attenuation contributed by the nitroxide probe, under magnetization equilibrium or off-equilibrium conditions.

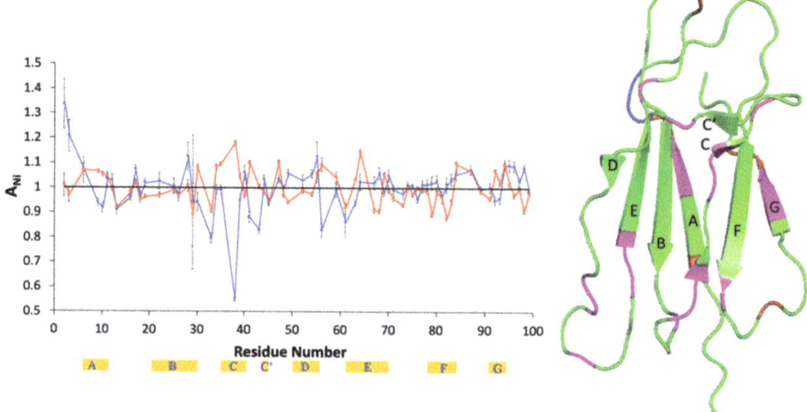

Figure 2. Overlay of the A_N values obtained from ^{1}H-^{15}N HSQC spectra of 8 μM β2m + 60 nM gold nanoparticles (AuNPs) in the presence of 0.8 mM Tempol, with a relaxation delay of 0.5 s (**blue**) or 5 s (**red**). The β-strand location and naming along the sequence is reported with yellow strips. The cartoon on the right highlights the positions of the accessible amides (**red**), i.e., exhibiting A_N[eq] > 1, and the amides with Type II deviation of A_N[off-eq] (**blue**), i.e., displaying A_N[off-eq] < A_N[eq]. The magenta color denotes sites where both A_N[eq] > 1 and Type II deviation occur simultaneously. The secondary structure elements of β2m are indicated according to the crystallographic naming scheme (PDB code 3HLA).

Table 3. Paramagnetic perturbation induced by 0.8 mM Tempol on the amide NMR signals of 8 μM β2m with 60 nM AuNPs. The equilibrium (column 2) and off-equilibrium (columns 3 and 4) data are reported in bold. The corresponding data obtained without AuNPs (Table 2) are reproduced below in plain fonts.

Structure Region	A_N > 1	Type I A_N [off-eq] > A_N [eq]	Type II A_N [off-eq] < A_N [eq]
N-term, strand A	**K6, V9, Y10, S11** V9	**Q2, R3** K6, Y10	**V9, Y10** Q2
loop AB	**N17**	**N17, K19** H13, E16, N17, K19	**R12, H13** R12
strand B	**F30** Y26	**F22, S28** C25, G29	**F30** Y26, S28, F30
loop BC	**D34** S33, D34		**S33, D34** D34
strand C	**I35, D38, L39, K41** I35, L39, K41	**L40** L40, K41	**I35, D38, L39, K41** I35, D38, L39
turn CC', strand C', loop C'D	**E47** G43, R45, E47	**E44, K48, V49** G43, K48, V49	**E47** R45, E47
strand D	**S55, F56** S52, L54, S55, F56	**S52, L54** L54	**S55, F56** S55, F56
loop DE	**D59** D59	S61	**D59**

Table 3. *Cont.*

Structure Region	$A_N > 1$	Type I A_N [off-eq] > A_N [eq]	Type II A_N [off-eq] < A_N [eq]
strand E	**Y63, L64, E69** L64, E69	**Y67, T68, F70** Y63, T68, F70	**Y63, L64, E69** L64, E69
loop EF	E77	**T71, T73, K75** T73, E74, K75, D76	E77
strand F	**H84** R81, V82, N83, H84	**A79, C80, V82, N83** T78	**H84** C80, R81, V82, N83
loop FG	**L87** Q89	L87	Q89
strand G, C-term	**I92, V93, K94, W95** I92, V93, R97, D98	**K91, W95, D96, R97, D98** V93, W95, D96, R97, D98, M99	**I92, V93** I92

A preliminary control of the ^{15}N longitudinal relaxation rates is useful to estimate the entities of the effects brought about by AuNPs and Tempol on β2m signals (Figure S4). Indeed, the average T_1 value of 8 μM β2m decreases by the same extent, i.e., 4%, with AuNPs or Tempol, whereas the addition of the nitroxide to β2m in AuNPs suspension shortens the average T_1 only by 0.7%. This means that the Tempol paramagnetic attenuation probing the statistical sampling differences would be extensively masked by the general and specific dipolar attenuation AuNPs inflict to β2m signals, if the A_N values were computed with respect to the isolated protein intensities. In conclusion, proper control intensities should be obtained from the protein sample in the presence of AuNPs rather than from the protein alone. Figure 2 shows the attenuation pattern of the backbone amide signals observed with 8 μM β2m and 60 nM AuNPs due to 0.8 mM Tempol. The numerical values are listed in Table S2, along with the corresponding errors. According to the preliminary T_1 control measurements, the A_N values are calculated with respect to the control solution, i.e., the sample with the same composition without the nitroxide. Compared to the A_N profile of the isolated protein (Figure 1), evident differences emerge in the A_N plot of Figure 2 concerning exposed or poorly accessible regions, as well as locations undergoing chemical or conformational exchange. In particular, AuNPs hinder accessibility at the N-terminal, end of strand B, loop CC′ and subsequent strand C′, start of strand D, strand F, and C-terminal of β2m, while increasing strand A exposure. The details of these differences can be appreciated from Table 3, which lists the equilibrium and off-equilibrium attenuation data with (bold fonts) and without (plain fonts) AuNPs, thereby enabling a direct comparison.

3. Discussion

The results above described must be analyzed from two viewpoints. On one side, the nitroxide-based screening on the free protein helps to gather elements on the association processes β2m undergoes in solution. These elements may reveal features related to the fibrillogenic propensity of β2m. On the other side, the same paramagnetic probing of the protein in the AuNPs suspension provides the elements that enable outlining the β2m interaction with those nanoparticles. Joining these two lines of evidence is particularly tempting, because it is thus possible to focus the mechanism of fibrillogenesis inhibition experimentally observed with AuNPs and the β2m variant D76N [4].

As pointed out in the previous section, the most relevant differences highlighted by equilibrium and off-equilibrium nitroxide collisional labeling when comparing the 8 μM and the 50 μM β2m solutions concern the higher exposure in the former of strands C, D, F, and G and the intensification in local conformational or chemical exchange processes at strands C and F, with a simultaneous loss of accessibility at strand A and loop AB (Table 2). This result is consistent with the previous inference and evidence addressing respectively the intermolecular interface in ΔN6 β2m, the variant devoid of the N-terminal hexapeptide fibrils [14], and the H-D exchanged β2m fibrils dissolved in DMSO [26]. In particular, it is very meaningful that the increased accessibility at strands C and F observed at lower concentration is paralleled by the onset, at the same positions, of exchange events (A_N[off-eq] Type II deviations, Table 2) that must witness the remnant of the intermolecular interaction propensity at those locations. At higher concentration, the pattern changes into significantly lower exposure and exchange due to the shift of the association dynamics toward oligomeric adducts.

The statistical sampling of Tempol shows that the presence of AuNPs exposes strand A and limits accessibility at the very N-terminal segment, end of strand B, turn CC′ and strand C′, start of strand D, strand F, and C-terminal region of β2m. Again, the gain or loss of accessibility at strands A, B, C′, and F, and turn CC′ are accompanied, at the same locations, by a gain or loss of Type II deviation of A_N[off-eq]. However, except for the N-terminal fragment (Q2, R3), the interaction pattern obtained by paramagnetic perturbation does not seem to match the previously reported picture based on relative intensity losses affecting essentially the end of strands B and D, and loops BC and DE [2]. The differences in β2m concentrations (8 vs. 26 μM) and protein/AuNPs ratios (100 vs. 200) between present and previous experiments could certainly play a role in determining some deviations. However, the relative intensity changes marking the difference between the absence and presence of AuNPs, and the paramagnetic perturbation measured in the presence of AuNPs, monitor different aspects of β2m interaction with AuNPs, and thus, the mismatch may be only apparent. In fact, similar to the previously reported low relative intensity of Q2 and R3 induced by AuNPs that translates into a Type I deviation of A_N[off-eq], i.e., hindered accessibility to the nitroxide probe, a similar effect with the paramagnetic perturbation is seen also for S28 (strand B, see Table 3). More frequently, instead, the β2m residues reported to lower the relative intensities due to AuNP interaction [2] exhibit Type II deviation of A_N[off-eq], i.e., local exchange, with nitroxide probing in the presence of AuNPs, which are typically coupled to above average accessibility (A_N[eq] > 1). This is the case with the ends of strands B and D, and with the involved subsequent locations at BC and DE loops (Table 3). Therefore, the picture emerging from the interpretation of the results obtained by equilibrium and off-equilibrium paramagnetic mapping does not conflict with the previous evidence but rather provides a more detailed and enriched characterization of the interaction between β2m and AuNPs. This appears quite clearly by the inspection of Figure 3 where the surface of the protein according to the equilibrium and off-equilibrium attenuation pattern induced by Tempol in the absence (upper structures) and presence (lower structures) of AuNPs is highlighted.

Hindered accessibility is marked by orange surfaces that change their distribution on moving from the isolated protein to the presence of AuNPs. In either conditions, those surfaces could be associated to regions that become screened by the relevant interaction, namely the residual protein-protein or the protein-nanoparticle one. The occurrence of exchange processes at the blue and magenta locations can be considered the consequence of an interaction that takes place over the ms-to-μs time scale and may represent a further interface with different dynamic properties with respect to the hindered accessibility surface, provided that the occurrence of a local conformational exchange process is ruled out [16]. Finally, the highly accessible positions that are identified in red indicate the surface that is not involved in any protein-protein nor protein-nanoparticle contact. Based on the discussed evidence for β2m alone [14,26] and with AuNPs [2,3], the results listed in Tables 2 and 3 and depicted in Figure 3 suggest that the AuNP interference leading to the inhibition of fibrillogenesis [4,6] could occur via interaction of the nanoparticles with the N-terminal and strands D and F of β2m, in addition to other contacts at the end of strand B, turn CC′, and strand C′. The suggested hypothesis is that these interactions in which AuNPs engage with the protein surface prevent the protein-protein contacts at the same locations that are necessary for fibrillogenic aggregation.

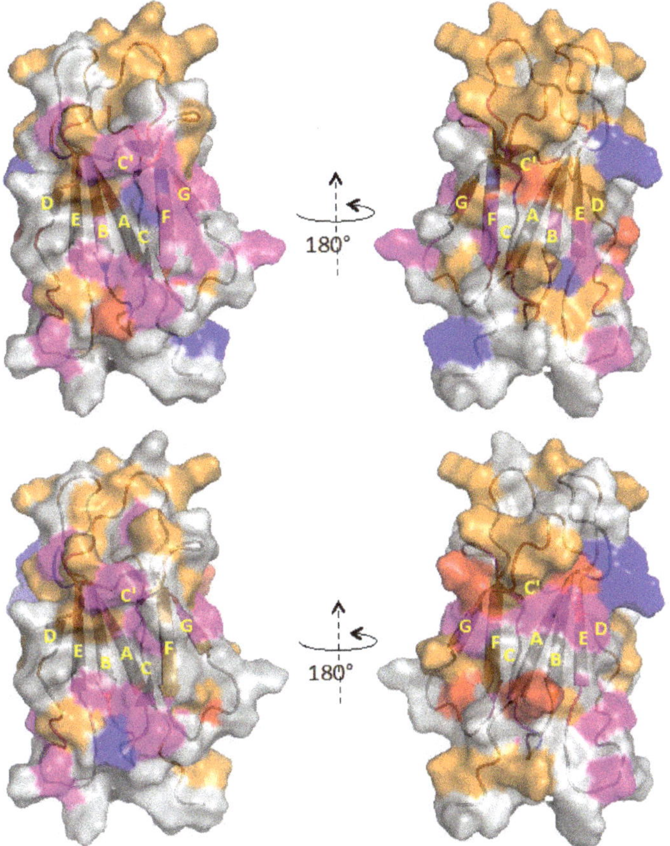

Figure 3. Cartoon representation of β2m surface sampled by Tempol in the absence (**upper pair**) and presence (**lower pair**) of AuNPs. The positions of the accessible backbone amides, i.e., exhibiting $A_N[\text{eq}] > 1$, are marked in red. The locations of the amides with Type I or Type II deviation of $A_N[\text{off-eq}]$, i.e., displaying $A_N[\text{off-eq}] > A_N[\text{eq}]$ or $A_N[\text{off-eq}] < A_N[\text{eq}]$, are highlighted in orange or blue, respectively. The magenta color denotes sites where both $A_N[\text{eq}] > 1$ and Type II deviation occur simultaneously. The very few positions where $A_N[\text{eq}] > 1$ and Type I deviation coincide were left in orange. The secondary structure elements of β2m are indicated according to the crystallographic naming scheme (PDB code 3HLA).

4. Materials and Methods

4.1. Chemicals

Sodium Citrate, ^{2}H$_2$O, Tempol (4-hydroxy-2,2,6,6-tetramethyl-piperidine-L-oxyl), and HEPES (*N*-(2-Hydroxyethyl)piperazine-*N'*-(2-ethanesulfonic acid) were all from Sigma Aldrich (St. Louis, MO, USA). From the same source were also the citrate-stabilized Au nanoparticles, here referred to as AuNPs. The average AuNP diameter was 5 nm, and the supplied suspension concentration was 91 nM.

4.2. Sample Preparation

The uniformly ^{15}N-labeled wild-type human β2m was expressed with an additional methionine at the N-terminus (Met-0) and purified as previously reported [14]. The protein samples in the absence of AuNPs were prepared in H$_2$O/D$_2$O 95/5, 1.5 mM sodium citrate, 20 mM HEPES buffer, pH 7.

The protein concentration was 8.0 µM, as determined by UV absorption at 280 nm. For solutions with AuNPs, proper amounts of D_2O and concentrated HEPES and β2m solutions were added to the mother NP suspension containing already citrate to reproduce the above-mentioned composition. Following dilution, the final AuNP concentration was 60 nM. A few microliters of concentrated Tempol solution were added when necessary to the NMR tube containing 0.550 mL of β2m, with or without AuNPs, to reach the desired Tempol/protein concentration ratio. For NMR samples, Tempol concentration was always 0.8 mM. For ESR samples, solutions at variable Tempol concentrations were prepared, i.e., 0.4, 0.8, and 1.6 mM, in aqueous buffer (20 mM HEPES, 1.5 mM sodium citrate, pH = 7), either alone or in the presence of 8 µM β2m, or 60 nM AuNP, or 8 µM β2m + 60 nM AuNP.

4.3. Spectroscopy

All the spectra were acquired at 298 K. The NMR experiments were collected at 14.0 T (^{1}H at 600.19 MHz, ^{15}N at 60.82 MHz) on a Bruker Avance III NMR system equipped with triple resonance cryoprobe. Two-dimensional ^{15}N-^{1}H HSQC experiments [27] carried out using sensitivity-improved Echo/Antiecho-TPPI pure phase detection in F1, gradient coherence selection, and flip-back pulse for solvent suppression [28–30] were acquired over spectral widths of 40 ppm and 14 ppm in F1 and F2 dimensions, respectively, with 64 time-domain points in t1, 256 or 512 scans × 2048 points in t2, and 64 dummy scans to achieve steady state. After a reproducibility check, relaxation delays were set to 0.5 and 5 s, respectively for off- and on-equilibrium conditions of magnetization recovery, following the guidelines previously reported [16]. The contour plots are reported in Figure S5, along with the corresponding signal-to-noise values (Table S3). The ^{15}N longitudinal relaxation times were measured using the sequence proposed by Kay and colleagues [31] with the modifications for sensitivity enhancement and flip-back pulse for solvent suppression [28–30]. The spectra with eight different relaxation intervals were acquired (10, 30, 60, 100, 140, 200, 400, and 1200 ms). All NMR data were processed with TOPSPIN version 4.0.2. Prior to Fourier transformation, linear prediction in t1 (up to 128 points) and zero filling were applied to yield a final data set of 2 K × 1 K points. For longitudinal relaxation analysis, the Bruker Dynamics Center 2.5.3 routine was used.

ESR spectroscopy experiments were collected with a Bruker EMXnano spectrometer operating in the X band. Capillaries filled with 50 µL of sample solution were placed in standard 4 mm tubes and submitted to acquisition (1 scan). The ESR operating parameters were as follows: frequency = 9.6 GHz; microwave power = 0.316 mW; modulation amplitude = 1 Gauss; modulation frequency = 100 kHz; center field = 3429.8 Gauss; sweep width = 200 Gauss; time constant = 1.28 ms. The data were processed using the software package Xenon (version 1.1b50, Bruker, Billerica, MA, USA).

4.4. Spectroscopic Data Treatment

Amide cross-peak intensities in ^{15}N-^{1}H HSQC spectra of β2m in the absence (I_d) and in the presence (I_p) of Tempol or/and AuNPs were measured by SPARKY software (version 3.133, T.D. Goddard and D.G. Kneller, University of California, San Francisco CA, USA). Normalized attenuation, A_N, was calculated according to Equation (1) [18]

$$A_N^k = \left(2 - \frac{\iota_p^k}{\iota_d^k} \right) \tag{1}$$

where the running index k refers to the kth residue amide cross-peak and the $\iota_{p,d}^k$ values are the corresponding auto-scaled intensities of the peaks in the presence (subscript p) and absence (subscript d) of nitroxide or/and AuNPs, which are defined as

$$\iota_{p,d}^k = \frac{I_{p,d}^k}{\frac{1}{n}\sum_{k=1}^{n} I_{p,d}^k} \tag{2}$$

with n representing the total number of measured peaks. From the above equation, it is seen that the scaling factor is simply the mean value over the n molecular locations for which the corresponding peak intensity can be estimated ($I^{av}_{p,d}$), the mean value of the individual auto-scaled intensities being unitary, by definition [16,18]. Therefore, values of A_N above or below unity indicate larger or smaller attenuations, respectively, with respect to the average absolute signal attenuation. From the definitions, the error on the individual A_N values can be calculated as [16]

$$\Delta A^k_N = A^k_N \times \sqrt{\left[\frac{\Delta I^k_p}{I^k_p}\right]^2 + \left[\frac{\Delta I^k_d}{I^k_d}\right]^2 + \left[\frac{\frac{1}{n^2}\sum\left(\Delta I^k_p\right)^2}{\left(I^{av}_p\right)^2}\right] + \left[\frac{\frac{1}{n^2}\sum\left(\Delta I^k_d\right)^2}{\left(I^{av}_d\right)^2}\right]} \tag{3}$$

where the first two terms under the square root sign represent the error on the relative intensity (*RI*) of the kth residue signal, i.e., the signal intensity ratio in the presence and absence of nitroxide or/and AuNPs, and the ΔI are the experimental intensity uncertainties obtained from the individual peak signal-to-noise figure.

The ESR spectra were employed to extract the rotational correlation time (τ_c) of Tempol in absence or presence of β_2m or/and AuNPs. Based on the method of Knowles and colleagues [32] and Kivelson's theoretical analysis [33], the τ_c values were estimated from:

$$\tau_c = 6.5 \times 10^{-10}\,\Delta B_0[(h_0/h_{-1})^{1/2-1}] \tag{4}$$

where ΔB_0 is the linewidth of the of the central line of the nitroxide ESR signal (a triplet because of the hyperfine coupling with the ^{14}N nuclear spin), and $h_{0/-1}$ are the amplitudes of the central and upfield lines. The corresponding error was calculated from Equation (4) by error propagation of the experimental uncertainties on linewidth ($\Delta\Delta B_0$) and amplitudes ($\Delta h_{0/-1}$), according to:

$$\Delta\tau_c = 6.5 \cdot 10^{-10}\left\{\left[\left(\frac{h_0}{h_{-1}}\right)^{\frac{1}{2}} - 1\right]\Delta\Delta B_0 + \frac{\Delta B_0}{2(h_{-1})^{\frac{1}{2}}(h_0)^{\frac{1}{2}}}\Delta h_0 + \frac{\Delta B_0(h_0)^{\frac{1}{2}}}{2(h_{-1})^{\frac{3}{2}}}\Delta h_{-1}\right\} \tag{5}$$

5. Conclusions

The results here described demonstrate that the paramagnetic perturbation methodology can be successfully applied to study protein-nanoparticle interactions. In addition to the surface accessibility mapping, extrinsic paramagnetic probes can provide valuable information on hindered accessibility and exchange processes by means of off-equilibrium attenuation analysis [16]. This methodology represents an additional tool that enriches the NMR relaxation approach to the characterization of protein interaction with the nanoparticle surface [34]. The delineation of the contact interface between protein monomers and between protein and nanoparticles is important not only for the comprehension of the mechanisms of protein aggregation and the elaboration of contrast strategies that bear particular relevance in amyloidogenic systems, but also for the characterization of that ensemble of labile contacts that is involved in the build-up of the so-called soft corona, i.e., the coating layer of weakly bound protein molecules with short residence times that can affect the nanoparticle targeting [35].

Supplementary Materials: The following are available online, Figure S1: overlay of ESR spectra; Figure S2: chemical shift perturbation values of β_2m amide signals under different conditions; Figure S3: relative intensity plots of 8 μM + 0.8 mM Tempol under magnetization equilibrium and off-equilibrium conditions; Figure S4: ^{15}N longitudinal relaxation times of β_2m amide signals under different conditions. Figure S5: ^{15}N-^{1}H HSQC maps; Table S1: A_N values plotted in Figure 1 and corresponding errors; Table S2: A_N values plotted in Figure 2 and corresponding errors; Table S3: individual peak signal-to-noise ratios of ^{15}N-^{1}H HSQC spectra.

Author Contributions: Conceptualization: G.E., Y.H., C.C., F.F.; Methodology: G.E., Y.H., C.C., F.F.; Formal analysis: Y.H., G.E., F.F.; Investigation: Y.H., G.E., F.F.; Resources: C.C., S.G., L.M.; Writing—Original Draft Preparation: G.E., Y.H.; Writing—Review and Editing: S.G., L.M., C.C., F.F., Y.H., G.E.; Supervision: G.E., F.F.;

Project Administration: G.E., Y.H., F.F.; Funding Acquisition: G.E. All authors have read and agreed to the published version of the manuscript.

Funding: This research was funded by NYUAD (grant No. 76 71260 ADHPG VP046).

Acknowledgments: We thank H. Molinari for stimulating the interest in the subject and the Core Technology Platform of NYUAD for the instrumentation. The assistance of Makek, A. is also gratefully acknowledged.

Conflicts of Interest: The authors declare no conflict of interest. The funder had no role in the design of the project, in the collection, analysis and interpretation of the data, in the writing of the paper, or in the decision to publish the results.

References

1. Kopp, M.; Kollenda, S.; Epple, M. Nanoparticle-Protein Interactions: Therapeutic Approaches and Supramolecular Chemistry. *Acc. Chem. Res.* **2017**, *50*, 1383–1390. [CrossRef]

2. Brancolini, G.; Corazza, A.; Vuano, M.; Fogolari, F.; Mimmi, M.C.; Bellotti, V.; Stoppini, M.; Corni, S.; Esposito, G. Probing the Influence of Citrate-Capped Gold Nanoparticles on an Amyloidogenic Protein. *ACS Nano* **2015**, *9*, 2600–2613. [CrossRef]

3. Brancolini, G.; Maschio, M.C.; Cantarutti, C.; Corazza, A.; Fogolari, F.; Corni, S.; Esposito, G. Citrate stabilized Gold Nanoparticles interfere with Amyloid Fibril formation: D76N and ΔN6 β2-microglobulin Variants. *Nanoscale* **2018**, *10*, 4793–4806. [CrossRef]

4. Cantarutti, C.; Raimondi, S.; Brancolini, G.; Corazza, A.; Giorgetti, S.; Ballico, M.; Zanini, S.; Palmisano, G.; Bertoncin, P.; Marchese, L.; et al. Citrate-stabilized Gold Nanoparticles hinder fibrillogenesis of a pathologic variant of β2-microglobulin. *Nanoscale* **2017**, *9*, 3941–3951. [CrossRef]

5. Cantarutti, C.; Bertoncin, P.; Corazza, A.; Giorgetti, S.; Mangione, P.P.; Bellotti, V.; Fogolari, F.; Esposito, G. Short-chain Alkanethiol Coating for Small-Size Gold Nanoparticles Supporting Protein Stability. *Magnetochemistry* **2017**, *3*, 40. [CrossRef]

6. Cantarutti, C.; Raj, G.; Fogolari, F.; Giorgetti, S.; Corazza, A.; Bellotti, V.; Naumov, P.; Esposito, G. Interference of citrate-stabilized gold nanoparticles on β2-microglobulin oligomeric association. *Chem. Commun.* **2018**, *54*, 5422–5425. [CrossRef]

7. Cantarutti, C.; Bertoncin, P.; Posocco, P.; Hunashal, Y.; Giorgetti, S.; Bellotti, V.; Fogolari, F.; Esposito, G. The interaction of β2-microglobulin and gold nanoparticles: Impact of coating, charge and size. *J. Mater. Chem. B* **2018**, *6*, 5964–5974. [CrossRef]

8. Gejyo, F.; Yamada, T.; Odani, S.; Nakagawa, Y.; Arakawa, M.; Kunitomo, T.; Kataoka, H.; Suzuki, M.; Hirasawa, Y.; Shirahama, T.; et al. A new form of amyloid protein associated with chronic hemodialysis was identified as β2-microglobulin. *Biochem. Biophys. Res. Commun.* **1985**, *129*, 701–706. [CrossRef]

9. Valleix, S.; Gillmore, J.D.; Bridoux, F.; Mangione, P.P.; Dogan, A.; Nedelec, B.; Boimard, M.; Touchard, G.; Goujon, J.M.; Lacombe, C.; et al. Hereditary Systemic Amyloidosis Dur to Asp76Asn Variant of β2-Microglobulin. *N. Engl. J. Med.* **2012**, *366*, 2276–2283. [CrossRef]

10. Verdone, G.; Corazza, A.; Viglino, P.; Pettirossi, F.; Giorgetti, S.; Mangione, P.; Andreola, A.; Stoppini, M.; Bellotti, V.; Esposito, G. The solution structure of human β2-microglobulin reveals the prodromes of its amyloid transition. *Protein Sci.* **2002**, *11*, 487–499. [CrossRef]

11. Esposito, G.; Corazza, A.; Viglino, P.; Verdone, G.; Pettirossi, F.; Fogolari, F.; Makek, A.; Giorgetti, S.; Mangione, P.; Stoppini, M.; et al. Solution structure of β2-microglobulin and insights into fibrillogenesis. *BBA Proteins Proteom.* **2005**, *1753*, 76–84. [CrossRef]

12. Rennella, E.; Corazza, A.; Giorgetti, S.; Fogolari, F.; Viglino, P.; Porcari, R.; Verga, L.; Stoppini, M.; Bellotti, V.; Esposito, G. Folding and Fibrillogenesis: Clues from β2-microglobulin. *J. Mol. Biol.* **2010**, *401*, 286–297. [CrossRef]

13. Mangione, P.P.; Esposito, G.; Relini, A.; Raimondi, S.; Porcari, R.; Giorgetti, S.; Corazza, A.; Fogolari, F.; Penco, A.; Goto, Y.; et al. Amyloid fibrillogenesis of Asp76Asn β2-microglobulin and co-aggregation with the wild type: The crucial roles of shear flow, natural hydrophobic surfaces and chaperones. *J. Biol. Chem.* **2013**, *288*, 30917–30930. [CrossRef] [PubMed]

14. Esposito, G.; Michelutti, R.; Verdone, G.; Viglino, P.; Hernandez, H.; Robinson, C.V.; Amoresano, A.; Dal Piaz, F.; Monti, M.; Pucci, P.; et al. Removal of the N-terminal hexapeptide from human β2-microglobulin facilitates protein aggregation and fibril formation. *Protein Sci.* **2000**, *9*, 831–845. [CrossRef]

15. Linse, S.; Cabaleiro-Lago, C.; Xue, W.F.; Lynch, I.; Lindman, S.; Thulin, E.; Radford, S.E.; Dawson, K.A. Nucleation of Protein Fibrillation by Nanoparticles. *Proc. Natl. Acad. Sci. USA* **2007**, *104*, 8691–8696. [CrossRef]

16. Hunashal, Y.; Cantarutti, C.; Giorgetti, S.; Marchese, L.; Molinari, H.; Niccolai, N.; Fogolari, F.; Esposito, G. Exploring exchange processes in proteins by paramagnetuc perturbation of NMR spectra. *Phys. Chem. Chem Phys.* **2020**, *22*, 6247–6259. [CrossRef]

17. Esposito, G.; Lesk, A.M.; Molinari, H.; Motta, A.; Niccolai, N.; Pastore, A. Probing protein structure by solvent perturbation of nuclear magnetic resonance spectra. *J. Mol. Biol.* **1992**, *224*, 659–670. [CrossRef]

18. Molinari, H.; Esposito, G.; Ragona, L.; Pegna, M.; Niccolai, N.; Brunne, R.M.; Lesk, A.M.; Zetta, L. Probing protein structure by solvent perturbation of NMR spectra: The surface accessibility of bovine pancreatic trypsin inhibitor. *Biophys. J.* **1997**, *73*, 382–396. [CrossRef]

19. Scarselli, M.; Bernini, A.; Segoni, C.; Molinari, H.; Esposito, G.; Lesk, A.M.; Laschi, F.; Temussi, P.A.; Niccolai, N. Tendamistat surface accessibility to the TEMPOL paramagnetic probe. *J. Biomol. NMR* **1999**, *15*, 125–133. [CrossRef]

20. Niccolai, N.; Ciutti, A.; Spiga, O.; Scarselli, M.; Bernini, A.; Bracci, L.; Di Maro, D.; Dalvit, C.; Molinari, H.; Esposito, G.; et al. NMR Studies of Protein Surface Accessibility. *J. Biol. Chem.* **2001**, *276*, 42455–42461. [CrossRef]

21. Benial, A.M.F.; Dhas, M.K.; Jawahar, A. Rotational Correlation Time Studies on Nitroxyl radicals Using 300 MHz ESR Spectrometer in High Viscous Liquid. *Appl. Magn. Reson.* **2011**, *40*, 311–319. [CrossRef]

22. Niccolai, N.; Bonci, A.; Rustici, M.; Scarselli, M.; Neri, P.; Esposito, G.; Mascagni, P.; Motta, A.; Molinari, H. NMR delineation of inner and outer protons from paramagnetic relaxation perturbations in 1D and 2D spectra of peptides. *J. Chem. Soc. Perkin Trans.* **1991**, 1453–1457. [CrossRef]

23. Giorgetti, S.; Raimondi, S.; Pagano, K.; Relini, A.; Bucciantini, M.; Corazza, A.; Fogolari, F.; Codutti, L.; Salmona, M.; Mangione, P.; et al. Effect of tetracyclines on the dynamics of formation and destructuration of β2-microglobulin amyloid fibrils. *J. Biol. Chem.* **2011**, *286*, 2121–2131. [CrossRef] [PubMed]

24. Esposito, G.; Garvey, M.; Alverdi, V.; Pettirossi, F.; Corazza, A.; Fogolari, F.; Polano, M.; Mangione, P.P.; Giorgetti, S.; Stoppini, M.; et al. Monitoring the interaction between β2-microglobulin and the molecular chaperone αB-crystallin by NMR and mass spectrometry. αB-Crystallin dissociates β2-microglobulin oligomers. *J. Biol. Chem.* **2013**, *288*, 17844–17858. [CrossRef] [PubMed]

25. Williamson, M.P. Using chemical shift perturbation to characterise ligand binding. *Prog. Nucl. Magn. Reson. Spectrosc.* **2013**, *73*, 1–16. [CrossRef] [PubMed]

26. Hoshino, V.; Katou, H.; Hagihara, Y.; Hasegawa, K.; Naiki, H.; Goto, Y. Mapping the core of of β2-microglobulin by H/D exchange. *Nat. Struct. Biol.* **2002**, *9*, 332–336. [CrossRef]

27. Bodenhausen, G.; Ruben, D.J. Natural abundance nitrogen-15 NMR by enhanced heteronuclear spectroscopy. *Chem. Phys. Lett.* **1980**, *69*, 185–189. [CrossRef]

28. Palmer, A.G.; Cavanagh, J.; Wright, P.E.; Rance, M. Sensitivity improvement in proton-detected two-dimensional heteronuclear correlation NMR spectroscopy. *J. Magn. Reson.* **1991**, *93*, 151–170. [CrossRef]

29. Schleucher, J.; Schwendinger, M.; Sattler, M.; Schmidt, P.; Schedletzky, O.; Glaser, S.J.; Sørensen, O.W.; Griesinger, C. A general enhancement scheme in heteronuclear multidimensional NMR employing pulsed field gradients. *J. Biomol. NMR* **1994**, *4*, 301–306. [CrossRef]

30. Grzesiek, S.; Bax, A. The Importance of Not Saturating H_2O in Protein NMR. Application to Sensitivity Enhancement and NOE Measurements. *J. Am. Chem. Soc.* **1993**, *115*, 12593–12594. [CrossRef]

31. Kay, L.E.; Torchia, D.E.; Bax, A. Backbone Dynamics of Proteins Studied by ^{15}N Inverse Detected Heteronuclear NMR Spectroscopy: Application to Staphylococcal Nuclease. *Biochemistry* **1989**, *28*, 8972–8979. [CrossRef] [PubMed]

32. Knowles, P.F.; Marsh, D.; Rattle, H.W.E. *Magnetic Resonance of Bio-Molecules: An Introduction to the Theory and Practice of NMR and ESR of Biological Systems*; Wiley: London, UK, 1976.

33. Kivelson, D. Theory of ESR Linewidths of Free Radicals. *J. Phys. Chem.* **1960**, *33*, 1094–1106. [CrossRef]

34. Ceccon, A.; Tugarinov, V.; Bax, A.; Clore, G.M. Global Dynamics and Exchange Kinetics of a Protein on the Surface of Nanoparticles Revealed by Relaxation-Based Solution NMR Spectroscopy. *J. Am. Chem. Soc.* **2016**, *138*, 5789–5792. [CrossRef] [PubMed]
35. Kumar, S.; Yadav, I.; Aswal, V.K.; Kohlbrecher, J. Structure and Interaction of Nanoparticle-Protein Complexes. *Langmuir* **2018**, *34*, 5679–5695. [CrossRef]

Sample Availability: Samples of Gold nanoparticles are commercially available from Sigma. Protein samples must be expressed and are currently not available from the authors.

Publisher's Note: MDPI stays neutral with regard to jurisdictional claims in published maps and institutional affiliations.

 *molecules*

Review

Protein Adsorption on Solid Supported Membranes: Monitoring the Transport Activity of P-Type ATPases

Francesco Tadini-Buoninsegni

Department of Chemistry "Ugo Schiff", University of Florence, 50019 Sesto Fiorentino, Italy; francesco.tadini@unifi.it; Tel.: +39-055-4573239

Academic Editor: Michael Assfalg
Received: 6 July 2020; Accepted: 2 September 2020; Published: 11 September 2020

Abstract: P-type ATPases are a large family of membrane transporters that are found in all forms of life. These enzymes couple ATP hydrolysis to the transport of various ions or phospholipids across cellular membranes, thereby generating and maintaining crucial electrochemical potential gradients. P-type ATPases have been studied by a variety of methods that have provided a wealth of information about the structure, function, and regulation of this class of enzymes. Among the many techniques used to investigate P-type ATPases, the electrical method based on solid supported membranes (SSM) was employed to investigate the transport mechanism of various ion pumps. In particular, the SSM method allows the direct measurement of charge movements generated by the ATPase following adsorption of the membrane-bound enzyme on the SSM surface and chemical activation by a substrate concentration jump. This kind of measurement was useful to identify electrogenic partial reactions and localize ion translocation in the reaction cycle of the membrane transporter. In the present review, we discuss how the SSM method has contributed to investigate some key features of the transport mechanism of P-type ATPases, with a special focus on sarcoplasmic reticulum Ca^{2+}-ATPase, mammalian Cu^+-ATPases (ATP7A and ATP7B), and phospholipid flippase ATP8A2.

Keywords: sarcoplasmic reticulum Ca^{2+}-ATPase; Cu^+-ATPase; phospholipid flippase; charge displacement; concentration jump; solid supported membrane; conformational transition; electrogenicity; ion translocation; phospholipid flipping

1. Introduction

P-type ATPases constitute a superfamily of membrane transporters that are present in all forms of life and are located in various membrane types, such as the plasma or cellular organelle membranes. The superfamily of P-type ATPases is classified into five distinct subfamilies (P1–P5), which are specific to different substrates [1–3]. These enzymes use the energy provided by ATP hydrolysis to transport various ions or phospholipids across cellular membranes, thereby generating and maintaining essential electrochemical potential gradients.

P-type ATPases share a similar molecular architecture, which comprises three distinct cytosolic domains, i.e., the actuator (A), nucleotide binding (N) and phosphorylation (P) domains, and two transmembrane domains, the transport domain of six helical segments (TM1 to TM6), which contains the ion binding sites located halfway through the membrane, and a class-specific support domain of four helical segments (TM7 to TM10). Moreover, in many P-type ATPases, the N- or C-terminal extensions at the cytosolic side act as regulatory (R) domains, which are autoinhibitory or can function as sensors for the transported cations [3,4]. Interestingly, the R domains of P-type ATPases have the characteristics of disordered proteins and are therefore highly variable and flexible. The disordered structure of the R domains is likely to facilitate their regulatory function favoring interaction with binding partners and helping to stabilize particular enzyme conformations [4,5].

P-type ATPases couple ion transport and ATP hydrolysis in a cyclic sequence of partial reactions that constitute the catalytic cycle. During catalysis, a transient phosphorylated intermediate is formed by the interaction of ATP with a conserved aspartate residue in the P domain, which is a specific feature of P-type ATPases. The Albers–Post or E_1–E_2 scheme [6,7] is the generally accepted model of the catalytic cycle of P-type ATPases. According to this model, the ATPase protein can assume two main conformational states, denoted E_1 and E_2, with different affinity for the transported ions and accessibility of the ion binding sites to the cytoplasmic and extracellular/luminal side. During the catalytic cycle, the ATPase undergoes structural rearrangements and conformational transitions between E_1 and E_2 states to perform ATP-driven transport of ions or phospholipids across the membrane [3,4].

The molecular mechanism of transport by P-type ATPases has been described in several reviews, see e.g., [1,3,4,8–10]. Figure 1 shows a simplified diagram of sequential reactions in the catalytic cycle of sarcoplasmic reticulum Ca^{2+}-ATPase (SERCA) [11]. Starting at the E_1 conformation, the SERCA cycle includes initial enzyme activation by high-affinity binding of two Ca^{2+} ions from the cytoplasmic side, followed by enzyme phosphorylation by ATP and the formation of a high-energy phosphorylated state $E_1{\sim}P$ (an ADP-sensitive phosphorylated intermediate that retains sufficient chemical energy to be able to transfer the phosphate to ADP, thus forming ATP). A conformational transition from the $E_1{\sim}P$ state to the lower energy phosphoenzyme intermediate E_2P (an ADP-insensitive phosphorylated intermediate whose relatively low energy is suggested by its non-reactivity with ADP) favors the translocation of Ca^{2+} ions across the membrane and their release into the sarcoplasmic reticulum (SR) lumen in exchange for two luminal protons. Hydrolytic cleavage of the phosphoenzyme E_2P (dephosphorylation) is followed by proton translocation and release to the cytosolic side, thus accelerating the E_2 to E_1 conformational transition, which completes the catalytic and transport cycle. Following the first high-resolution crystal structure of SERCA with bound Ca^{2+} (i.e., the $E_1{\cdot}Ca_2$ state) [12], several crystal structures of SERCA in different conformational states in the transport cycle have been determined at atomic resolution, as reviewed in e.g., [10,13–19].

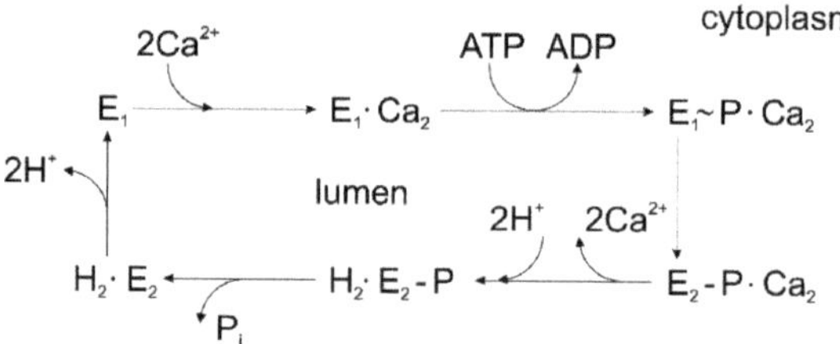

Figure 1. Simplified diagram of sequential reactions in the transport cycle of sarcoplasmic reticulum Ca^{2+}-ATPase (SERCA).

An electrophysiological method based on solid supported membranes (SSMs) has been used successfully to monitor the transport of charged substrates in various membrane transporters, including P-type ATPases [20]. SSM measurements of electrical currents can provide mechanistic and kinetic information about the movement of charged substrates within the membrane transporter as well as about conformational transitions associated with charge transfer in the reaction cycle of the membrane transport protein. In the present review, we will present the main features of the SSM-based electrophysiological method and discuss how the technique has contributed to investigating key aspects of the transport mechanism of P-type ATPases, with a special focus on SERCA, mammalian Cu^+-ATPases, and a phospholipid flippase.

2. Current Measurements on Solid Supported Membranes

The SSM represents a convenient model system for a lipid bilayer membrane. In particular, the SSM consists of a hybrid alkanethiol/phospholipid bilayer supported by a gold electrode. The SSM is formed by covering the gold surface with an alkanethiol monolayer, usually an octadecanethiol monolayer, and then by self-assembling a phospholipid monolayer on top of the gold-supported thiol layer [21,22]. The so-formed hybrid bilayer (Figure 2) is characterized by a high mechanical stability so that fast solution exchange can be performed at the SSM surface. The exchange of solutions provides the substrate or ligand and activates the membrane transporter adsorbed on the SSM [22].

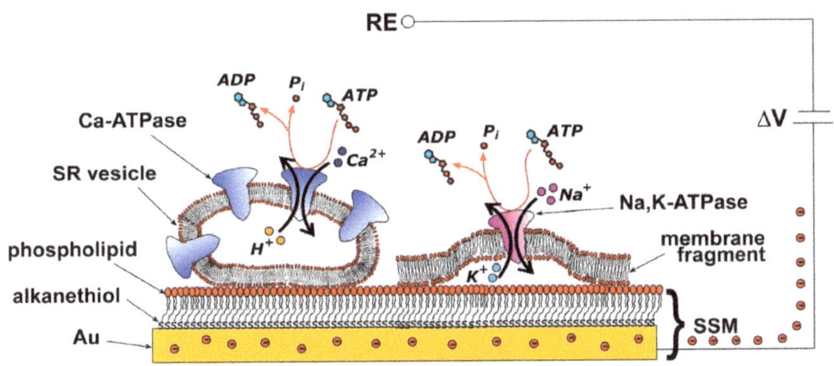

Figure 2. Schematic diagram of an sarcoplasmic reticulum (SR) vesicle containing Ca^{2+}-ATPase and of a membrane fragment incorporating Na^+,K^+-ATPase adsorbed on a solid supported membranes (SSM) and subjected to an ATP concentration jump. If the ATP jump induces charge movement across the ATPase, a compensating electrical current flows along the external circuit (the red spheres represent electrons) if the potential difference (ΔV) applied across the whole system is kept constant. RE, reference electrode. Adapted with permission from [23]. Copyright 2009 American Chemical Society.

Various membrane preparations containing the transport protein of interest, i.e., native membrane vesicles, purified membrane fragments, and proteoliposomes with reconstituted proteins, can be physically adsorbed on the SSM (Figure 2). Adsorption of such membrane preparations allows a variety of transport proteins to be immobilized on the SSM surface in a simple spontaneous process. This experimental approach is much easier and more effective than direct incorporation of the membrane transporter in a free-standing planar lipid bilayer, such as the black lipid membrane, which requires complicated incorporation procedures, leading to a superior signal-to-noise ratio and time resolution of the electrical measurement.

Following stable adsorption of the membrane sample on the SSM, the membrane transporter is subjected to a substrate concentration jump through the solution exchange technique. A rapid exchange from a solution with no substrate for the membrane transporter to one containing a specific substrate, e.g., ATP for P-type ATPases, activates the transport protein. If the substrate concentration jump induces charge displacement across the protein, an electrical current is measured due to capacitive coupling between the membrane sample and the SSM [20,24,25]. In particular, movement of a net charge across the activated protein is compensated by a flow of electrons along the external circuit toward the electrode surface, to keep constant the potential difference (ΔV) applied across the whole metal/solution interphase (Figure 2) [20]. This flow of electrons corresponds to the measured capacitive current, which is strictly correlated with the transporter-generated current and is recorded as a transient current signal [20,24,25]. The SSM method allows the measurement of charge displacement under pre-steady state conditions, while steady-state currents are not recorded. We point out that the electrical behavior of the system is essentially the same whether membrane vesicles or membrane fragments are adsorbed on the SSM.

The transport mechanism of various P-type ATPases belonging to different subfamilies was characterized using the SSM technique, such as in the case of Na^+,K^+-ATPase [22,26], SERCA [27,28], and H^+,K^+-ATPase [29], belonging to the P2-ATPase subfamily, and more recently bacterial and mammalian Cu^+-ATPases of subclass P1B [30,31] and P4-ATPase phospholipid flippase [32]. On the other hand, the P3-ATPase subfamily, which comprises plasma membrane H^+-ATPases of fungal and plant cells, has not yet been investigated by the SSM method.

SSM measurements on P-type ATPases were useful to identify electrogenic steps, i.e., reaction steps associated with a net charge transfer, and to assign time constants to partial reactions in the ATPase transport cycle. However, slow transport processes with time constants greater than 200 ms can be hardly recorded in SSM-based current measurements [20].

Finally, the SSM technique has been successfully employed to evaluate the effects of pharmacologically relevant compounds, such as anti-cancer drugs [33], on the transport activity of P-type ATPases and to characterize the interaction of specific ATPase inhibitors, thereby providing a quantitative estimate of inhibition potency (IC_{50} values).

Analysis systems for SSM-based electrophysiology are commercially available and are based on the SURFE^{2}R (Surface Electrogenic Event Reader) technology, as described in [29,34–36]. When higher throughput is required as in the case of drug screening, a fully automated device allows measuring electrical currents simultaneously from 96 individual SSM sensors in a parallel mode.

3. P-Type ATPases Investigated on Solid Supported Membranes

As mentioned above, the SSM technique was used to investigate net charge translocation (electrogenic transport) in P-type ATPases. Charge displacement associated with specific steps, i.e., ion binding/release, ion translocation, and exchange was measured in the ATPase transport cycle and the electrogenicity of partial reactions was determined, thereby providing mechanistic insights in the transport mechanism of different P-type ATPases. For example, a direct proof for the electrogenicity of cytoplasmic Na^+ binding to the Na^+,K^+-ATPase was obtained with Na^+ concentration jump experiments performed on membrane fragments containing Na^+,K^+-ATPase adsorbed on the SSM [26]. It was found that the charge associated with the Na^+ binding step is about 30% of the displaced charge related to Na^+ translocation and release, indicating that cytoplasmic Na^+ binding is a minor electrogenic event in the reaction cycle of Na^+,K^+-ATPase [26].

In the next sections, we will discuss the contribution of the SSM technique to unravel key features of the electrogenic transport activity of some prominent members of the P-type ATPase family. In particular, the focus of the present review is on SERCA, Cu^+-ATPases ATP7A and ATP7B, and P4-ATPase (phospholipid flippase) ATP8A2. SERCA has been characterized in detail by the SSM technique, providing useful information on the enzyme's transport mechanism. This information was used for a comparative analysis of the transport properties of the Cu^+-ATPases and phospholipid flippase, which were recently investigated by the SSM method.

3.1. Sarcoplasmic Reticulum Ca^{2+}-ATPase

The SERCA enzyme is one of the most investigated P-type ATPase [15,16,37–39]. In muscle cells, SERCA couples the energy gained by the hydrolysis of one ATP molecule to the transport of two Ca^{2+} ions against their electrochemical potential gradient from the cytoplasm into the lumen of SR, which is the main intracellular Ca^{2+} storage organelle. Ca^{2+} uptake in the SR lumen by SERCA plays an essential role in regulating cytoplasmic Ca^{2+} concentration, which is kept at or below 0.1 μM; in this manner, SERCA induces muscle relaxation and contributes to intracellular Ca^{2+} homeostasis. Modified SERCA expression and impaired pumping activity have been associated with pathological conditions and several diseases with a wide range of severity [39,40].

SERCA (approximately 110 KDa) belongs to the P2A-ATPase subfamily. In mammals, SERCA is encoded by three different genes, ATP2A1-3, but isoform diversity is increased by alternative splicing of the transcripts, which raises the number of possible SERCA isoforms to more than 10 [41,42]. A very convenient experimental system for functional and structural studies of SERCA is provided by vesicular fragments of longitudinal SR, where SERCA1a is the predominant isoform. SR vesicles contain a high amount of SERCA, which accounts for approximately 50% of the total protein and which reaches a density in the SR membrane of about 30,000 μm^{-2} [43].

Electrical currents generated by SERCA were measured by adsorbing native SR vesicles containing SERCA1a from rabbit skeletal muscle on the SSM and by activating the calcium pumps with substrate, i.e., Ca^{2+} and ATP concentration jumps. The observed current signals allow the direct measurement of charge translocation by SERCA under different activation conditions. In particular, charge movements related to different electrogenic partial reactions in the SERCA transport cycle were detected. It was shown that a Ca^{2+} concentration jump in the absence of ATP induces a transient current (dotted line in Figure 3A), which is associated with an electrogenic event corresponding to enzyme activation by the initial binding of Ca^{2+} to the cytoplasmic side of the ATPase (the exterior of the SR vesicle, see Figure 2) [27,28,44]. When an ATP concentration jump was performed in the presence of Ca^{2+} ions, a current signal was detected (solid line in Figure 3A), which is associated with a further electrogenic step corresponding to ATP-dependent calcium translocation by the enzyme [20,27]. In particular, ATP concentration jump experiments on SR vesicles in the presence and absence of a calcium ionophore at different pH values [27] indicated that the ATP-induced electrical current is related to displacement and release of pre-bound Ca^{2+} at the luminal side of the pump (the interior of the SR vesicle, see Figure 2) after phosphorylation of the enzyme by ATP. The transient currents measured after a Ca^{2+} jump in the absence of ATP and an ATP jump in the presence of Ca^{2+} were both fully suppressed by thapsigargin [44], which is a highly specific and potent SERCA inhibitor [45,46]. We point out that to perform ATP hydrolysis and active Ca^{2+} transport SERCA undergoes large domain movements enabled by dynamic fluctuations and conformational transitions that are not random but instead are driven by the availability of specific substrates [47].

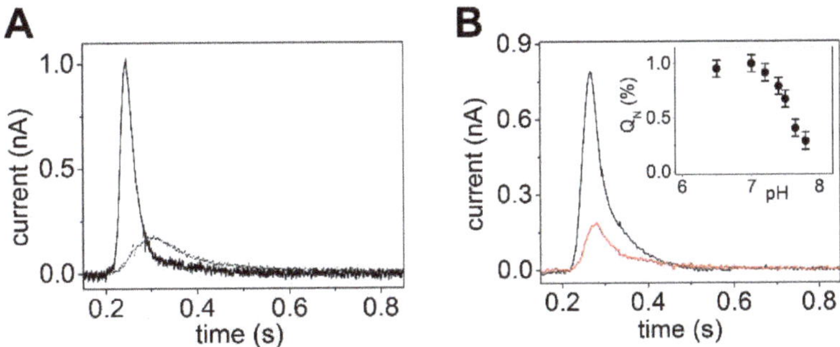

Figure 3. Transient currents generated by SERCA adsorbed on an SSM. (**A**) Transient current after a 10 μM free Ca^{2+} concentration jump in the absence of ATP (dotted line) and a 100 μM ATP concentration jump in the presence of 10 μM free Ca^{2+} (solid line). Reprinted from [44] with permission. (**B**) Current signals after 100 μM ATP concentration jumps in the presence of 10 μM free Ca^{2+} and 100 mM KCl at pH 7 (black line) and 7.8 (red line). The inset shows the dependence of the normalized charge (Q_N) after 100 μM ATP concentration jumps on pH. The charges were normalized with respect to the maximum charge measured at pH 7. S.E. are given by error bars. Adapted from [48].

It is interesting to observe that the amplitude of the signal related to ATP-dependent Ca^{2+} translocation decreases as the pH is raised from 7 to 8 (Figure 3B). It is known that exchange of Ca^{2+} with H^+ is a specific feature of SERCA [37,48], which favors Ca^{2+} release at the luminal side [17,49]. Useful information was provided by previous measurements on reconstituted proteoliposomes containing SERCA [49–52]. In particular, it was shown that the stoichiometry of the Ca^{2+}/H^+ countertransport is about 1/1 when the luminal and medium pH is near neutrality [49,52]. The importance of Ca^{2+}/H^+ exchange in determining the release of bound Ca^{2+} from the phosphoenzyme E_2P was demonstrated in steady-state experiments on native SR vesicles [37]. It was reported that the maximal levels of accumulated Ca^{2+} are significantly reduced if the pH is raised above 7. This result shows that if exchange is limited due to low H^+ concentration, Ca^{2+} is less likely to dissociate from the phosphoenzyme. Thus, the pH dependence of the current signals obtained with ATP concentration jumps (inset of Figure 3B) also indicates that when a lack of H^+ limits Ca^{2+}/H^+ exchange, i.e., alkaline pH, the translocation of bound Ca^{2+} is prevented, even though K^+ is present in high concentration and may neutralize acid residues at alkaline pH [48]. This suggests a requirement for specific H^+ binding at the Ca^{2+} transport sites in order to obtain Ca^{2+} release.

The SSM method has also been used to investigate a very interesting research topic, which is currently receiving much attention, i.e., the molecular mechanisms of SERCA regulation. In muscle cells, SERCA transport activity is regulated by two analogous transmembrane proteins: phospholamban (PLN, 52 amino acids), which is primarily expressed in cardiac muscle where it regulates the SERCA2a isoform [53], and sarcolipin (SLN, 31 amino acids), which is mainly expressed in skeletal muscle where it regulates the SERCA1a isoform [54]. In particular, PLN inhibits pump activity by lowering the apparent Ca^{2+} affinity of SERCA, and the phosphorylation of PLN by protein kinases relieves SERCA inhibition [53]. There is general consensus that the PLN inhibition of SERCA involves the reversible physical interaction of a PLN monomer under calcium-free conditions. However, experimental evidence was provided that a PLN pentamer, which has been described as an inactive storage form, can also interact with SERCA [55,56].

To investigate the PLN effect on ATP-dependent Ca^{2+} translocation by SERCA, SSM-based current measurements were carried out on co-reconstituted proteoliposomes containing SERCA and PLN [57]. The proteoliposomes were adsorbed on the SSM and activated by Ca^{2+} and/or ATP concentration jumps. In particular, substrate conditions (various Ca^{2+} and ATP concentrations) were chosen that promoted specific conformational states of SERCA, from which calcium transport could be initiated. The results from pre-steady state charge (calcium) translocation experiments were compared with steady-state measurements of ATPase hydrolytic activity. It was found that the PLN effect on SERCA transport activity depends on substrate conditions, and PLN can establish an inhibitory interaction with multiple conformational states of SERCA (a calcium-free E_2 state, a E_1-like state promoted by Ca^{2+}, and a E_2-like state promoted by ATP, shown in red in Figure 4) with distinct effects on SERCA's kinetic properties [57]. It was also noted that once a particular SERCA–PLN inhibitory interaction is established, it remains throughout the SERCA transport and catalytic cycle. These findings were interpreted on the basis of a conformational memory [58,59] in the interaction of PLN with SERCA, whereby a defined structural state of the SERCA/PLN regulatory complex, which depends on substrate conditions, is retained during SERCA turnover and conformational cycling.

In addition to PLN and SLN, single-span transmembrane proteins have recently been discovered that act as regulators of SERCA activity: dwarf open reading frame (DWORF), myoregulin (MLN), endoregulin (ELN), and another-regulin (ALN) [60–62]. While MLN, ELN, and ALN have been identified as inhibitors of SERCA activity, it was shown that DWORF does not inhibit the SERCA pump [62], enhancing Ca^{2+} uptake by displacing PLN. The oligomerization of these new SERCA regulators and the binding interaction of the monomeric form with the calcium pump were very recently investigated [63], thus providing a useful contribution in the characterization of the complexity of SERCA regulatory mechanisms. In this respect, it appears that the above-mentioned transmembrane peptides could be conveniently investigated by the SSM technique upon their reconstitution in proteoliposomes containing SERCA. This would help to elucidate the inhibitory or activation effects of the recently discovered SERCA regulators.

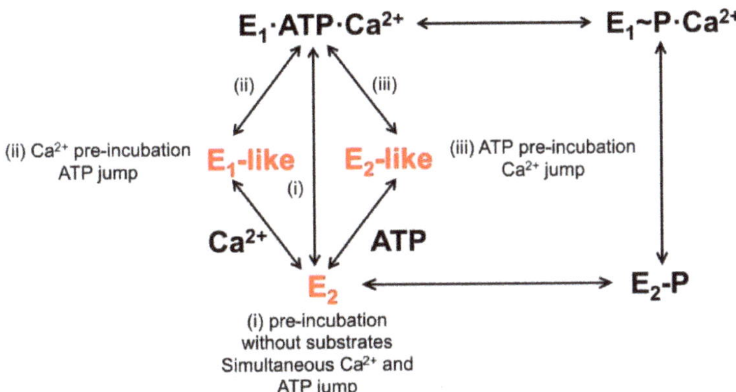

Figure 4. The SERCA transport cycle with relevant conformational states. The pre-incubation and concentration jump conditions used [57] are indicated. Shown in red are the calcium-free E$_2$ state, an E$_1$-like state promoted by calcium, and an E$_2$-like state promoted by ATP. Adapted from [57].

3.2. Cu$^+$-ATPases ATP7A and ATP7B

The mammalian copper ATPases ATP7A and ATP7B are 165–170 KDa membrane proteins belonging to subclass IB of the P-type ATPase superfamily. At normal copper levels in the cell, ATP7A and ATP7B are found in the trans-Golgi network (TGN), and these enzymes translocate copper across the membrane from the cytoplasm into the TGN lumen using ATP hydrolysis [64–68]. ATP7A and ATP7B contribute to intracellular copper homeostasis by delivering copper to newly synthesized copper-containing proteins in the TGN and by removing copper excess from the cell [64]. ATP7A is expressed in most tissues but not in the liver, whereas ATP7B is mainly found in this organ [64]. The malfunction of either ATP7A or ATP7B is the cause of severe diseases, which are known as Menkes (ATP7A) and Wilson (ATP7B) diseases.

ATP7A and ATP7B show high sequence homology (about 60% identity). Their structure comprises eight transmembrane helices, which include a copper binding site (transmembrane metal binding site, TMBS), and the A, N and P cytoplasmic domains, which are common for P-type ATPases. A unique structural feature of ATP7A and ATP7B is the highly mobile N-terminal chain of six copper binding domains (N-terminal metal binding domain) that are involved in the copper-dependent regulation and intracellular localization of these enzymes [69].

As described in several reviews (e.g., [64,65,70–74]), Cu$^+$ transfer by ATP7A and ATP7B involves copper acquisition from donor proteins on the cytoplasmic side of the membrane and copper delivery to acceptor proteins on the luminal side, without establishing a free Cu$^+$ gradient. In conformity with other P-type ATPases, ATP7A and ATP7B hydrolyze ATP to form a transient phosphorylated intermediate, and they undergo conformational transitions that favor Cu$^+$ transfer to/from the TMBS. From high-resolution crystal structures and molecular dynamics simulations on a bacterial Cu$^+$-ATPase (*Legionella pneumophila* Cu$^+$-ATPase, LpCopA) [75,76], it appears that copper ATPases have a unique copper release mechanism that is likely to be involved in specific and controlled Cu$^+$ delivery to acceptor proteins.

Electrogenic copper movement within mammalian copper ATPases was demonstrated by current measurements on COS-1 microsomes expressing recombinant Cu$^+$-ATPases (ATP7A and ATP7B) adsorbed on an SSM [31,48]. When an ATP concentration jump was performed on microsomes containing ATP7B (or ATP7A) in the presence of CuCl$_2$ and dithiothreitol to reduce Cu^{2+} to

Cu$^+$, a current signal was obtained (solid line in Figure 5A), which was not observed when bathocuproinedisulfonate (BCS), acting as Cu$^+$ chelator, was added to the reaction buffer (dotted line in Figure 5A). These experiments indicate that the copper-related current signal is associated with an electrogenic event corresponding to Cu$^+$ movement within ATP7B upon phosphorylation by ATP [31,48], which is consistent with copper displacement from the TMBS to the luminal side of the enzyme.

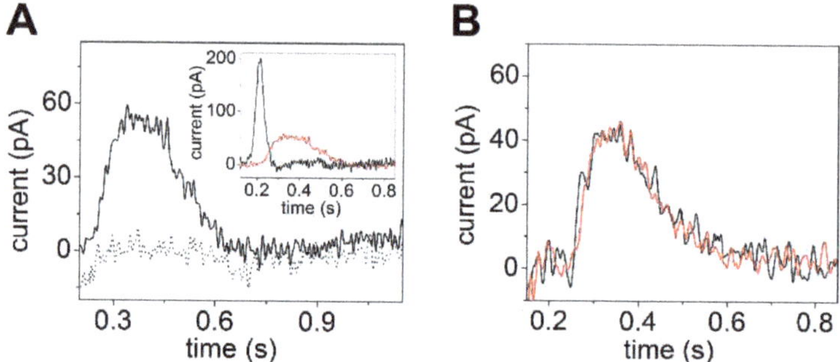

Figure 5. Transient currents generated by ATP7B adsorbed on an SSM. (**A**) Transient currents after 100 µM ATP concentration jumps in the presence of 5 µM CuCl$_2$ (solid line) or 1 mM bathocuproinedisulfonate (BCS) (dotted line). The inset shows current signals after 100 µM ATP concentration jumps on ATP7B (red line) and SERCA (black line). Adapted from [31] with permission from Wiley. (**B**) Current signals after 100 µM ATP concentration jumps in the presence of 5 µM CuCl$_2$ at pH 6 (black line) and 7.8 (red line). Adapted from [48].

By fitting the decay phase of the transient current with a first-order exponential decay function, a charge transfer decay time constant (τ) of 140 ms was determined for ATP7B, which is within the time frame of aspartate phosphorylation by ATP [31], suggesting that copper displacement in ATP7B is correlated to formation of the phosphorylated intermediate and precedes phosphoenzyme hydrolytic cleavage. This conclusion was also supported by SSM-based current measurements on the D1044A mutant of ATP7A. Asp1044 is the conserved aspartate residue in the P-domain of ATP7A that interacts with ATP to form the aspartyl phosphorylated intermediate. It was shown that the D1044A mutant yielded no current signal upon an ATP concentration jump in the presence of Cu$^+$ [77]. This result further indicated that ATP-dependent copper movement through the ATPase is directly correlated to formation of the aspartyl phosphorylated intermediate by ATP consumption.

It is interesting to observe that ATP-induced copper movement in mammalian Cu$^+$-ATPases is significantly slower than ATP-dependent Ca^{2+} translocation in SERCA [31], as shown by the different decay time constants τ for charge displacement following ATP jumps (inset of Figure 5A) on ATP7B (red line, $\tau = 140$ ms) and SERCA (black line, $\tau = 25$ ms). It is worth mentioning that the τ values for charge movements in ATP7B and SERCA are consistent with a slower phosphoenzyme formation in the copper ATPase [31] with respect to SERCA [78]. It should be noted that these decay time constants are attributed to initial partial reactions of the pump transport cycle and are not equivalent to steady-state turnover [31].

SSM measurements on ATP7A and ATP7B revealed that ATP-induced charge movement in these enzymes is not changed by alkaline or acid pH [48], as shown by charge transfer measurements at different pH values (Figure 5B). This finding indicated that copper displacement in ATP7A and ATP7B is pH independent, and it highlights a significant difference in the transport mechanisms of ATP7A/B and SERCA. It was proposed that in ATP7A/B, Cu$^+$/H$^+$ exchange may not be required for luminal copper release [48], as opposed to the strict requirement of Ca^{2+}/H$^+$ exchange in SERCA as discussed above. It is worth mentioning that carboxylate residues are absent in the ion-binding cluster located in the transmembrane region of the bacterial *Archaeoglobus fulgidus* CopA [79] and LpCopA [30], while crucial aspartate and glutamate residues are present in the equivalent transmembrane domain of SERCA [12,16,80] that are directly involved in Ca^{2+}/H^+ exchange. Thus, SSM measurements on ATP7A/B supported the hypothesis that Cu^+ release in these enzymes may not be coupled to a net proton countertransport, which has not been observed for PIB-type ATPases [72,73,81]. Interestingly, a very recent study reported real-time fluorescence measurements on *E.coli* Cu^+-ATPase (EcCopA) reconstituted in small unilamellar vesicles encapsulating a set of fluorescence probes that are selective for Cu^+, pH, and membrane potential [82]. The results of this study demonstrated the absence of H^+ countertransport in the Cu^+ translocation cycle of EcCopA, qualifying EcCopA as an electrogenic uniporter.

3.3. P4-ATPase ATP8A2

A characteristic feature of eukaryotic cell membranes is the asymmetrical distribution of different lipids across the membrane bilayer. This is particularly evident in the plasma membrane, where phosphatidylcholine (PC) and sphingolipids, i.e., sphingomyelin and glycosphingolipids, are concentrated in the exoplasmic leaflet of the membrane, whereas phosphatidylserine (PS) and phosphatidylethanolamine (PE) are mainly restricted in the cytoplasmic leaflet [83–86]. Membrane lipid asymmetry is essential for a variety of cellular processes that include, e.g., cell and organelle shape determination, membrane stability and impermeability, membrane curvature, vesicle formation and trafficking, host–virus interactions, membrane protein regulation, blood coagulation, and apoptosis [86–90].

Phospholipid flippases, belonging to the P4-ATPase subfamily, couple ATP hydrolysis to the translocation of specific phospholipids from the exoplasmic to the cytoplasmic leaflet of biological membranes in order to generate and maintain transmembrane lipid asymmetry [89,91–95]. P4-ATPases are only found in eukaryotes and constitute the largest P-type ATPase subfamily. In mammals, at least 14 P4-ATPases are known, which are divided into five classes [89]. P4-ATPase dysfunction has been associated with severe neurological disorders and liver diseases in humans [92]. These lipid transporters consist of a large polypeptide with a molecular mass of about 120 kDa, which shares the general architecture of P-type ATPases. Most P4-ATPases form a heterodimeric complex with an accessory β-subunit of about 50 kDa belonging to the CDC50/LEM3 family [89,96,97]. High-resolution structures of yeast [98,99] and human [100] lipid flippases were determined by cryo-electron microscopy, as reviewed in [101], and very recently, the crystal structures of a human plasma membrane flippase were also reported [102].

The transport mechanism of P4-ATPases is the subject of intensive research, and various models have been proposed for the phospholipid translocation pathway in P4-ATPases [103–107]. The recent atomic resolution structures of yeast and human P4-ATPases have provided valuable information on different conformational states in the flippase transport cycle, which is depicted by the Albers–Post or E_1–E_2 scheme commonly used to describe the mechanism of ion transporting P2-type ATPases. The P4-ATPase reaction cycle (see the simplified diagram in Figure 6A) has been examined in some detail for the mammalian flippase ATP8A2 [108], which is highly expressed in the retina, brain, testis, and spinal cord. It was shown that ATP8A2 is phosphorylated by ATP at the aspartate conserved in all P-type ATPases, and the phosphoenzyme exists in E_1P and E_2P states [108]. Dephosphorylation of the E_2P state is activated by binding of the two known substrates PS and PE, but not by binding of PC that is not a substrate of ATP8A2 [109], and dephosphorylation is associated with lipid translocation from the exoplasmic to the cytoplasmic leaflet of the membrane bilayer. Although significant progress has been made in our understanding of phospholipid flipping by P4-ATPases, several aspects of the flippase transport mechanism remain to be explored, such as the stoichiometry of phospholipid molecules translocated per ATP hydrolyzed, the mechanisms underlying lipid binding and release, the electrogenicity of phospholipid transport, and the related issue of countertransport, i.e., countertransport of an ion or other charged substrate from the cytoplasm to the exoplasm in connection with the $E_1 \rightarrow E_1P \rightarrow E_2P$ reaction sequence as observed for P2-type ATPases.

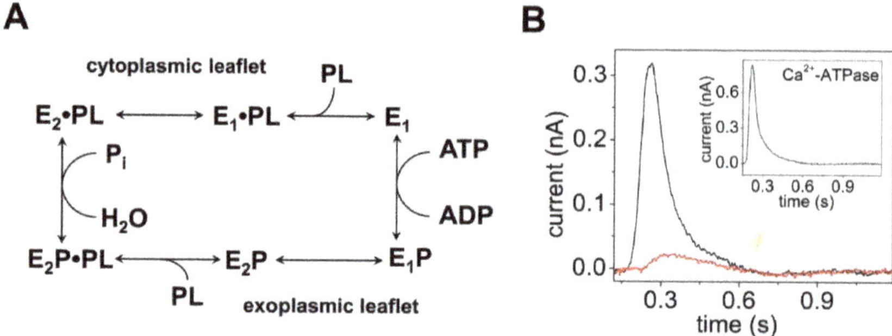

Figure 6. Simplified diagram of the ATP8A2 reaction cycle and transient currents generated by ATP8A2 adsorbed on a SSM. (**A**) E_1, E_1P, E_2P, and E_2 are different enzyme conformational states, where "P" indicates covalently bound phosphate. The phospholipid substrate, PL (phosphatidylserine (PS) or phosphatidylethanolamine (PE)), enters the cycle from the exoplasmic leaflet of the lipid bilayer by binding to the E_2P phosphoenzyme, thereby stimulating the dephosphorylation and release of the lipid substrate toward the cytoplasmic leaflet as a consequence of the E_2 to E_1 conformational change. (**B**) Current transients observed upon 100 µM ATP concentration jumps on ATP8A2 reconstituted in proteoliposomes containing a mixture of 90% PC and 10% PS, in the absence (black line) or in the presence (red line) of 50 µM orthovanadate. The inset shows the current signal induced by a 100 µM ATP concentration jump on native SR vesicles containing SERCA. Reprinted from [32].

To address unexplored key aspects of the flipping mechanism of P4-ATPases, in particular the electrogenicity of phospholipid flippases and ion countertransport, the SSM method was very recently used in a study of the elctrogenic properties of wild-type and mutant forms of the flippase ATP8A2 [32]. Purified ATP8A2 and its accessory CDC50A protein were reconstituted in proteoliposomes of different lipid compositions that were adsorbed on the SSM surface and subjected to ATP concentration jumps. It was shown that an ATP jump on ATP8A2 reconstituted into proteoliposomes consisting of a mixture of 90% PC and 10% PS induced a current signal (black line in Figure 6B) that was completely suppressed in the presence of the ATPase inhibitor orthovanadate (red line Figure 6B). Since orthovanadate binds to the ATPase from the cytoplasmic side, it was concluded that the ATPase molecules with the cytoplasmic side facing the external aqueous solution generated the ATP-dependent charge movement across ATP8A2. It was also noted that the sign of the ATP8A2-related current signal is positive, as observed for the SERCA-related transient current (inset of Figure 6B) that is attributed to the translocation and release of Ca^{2+} ions into the SR vesicle interior [27] (see Section 3.1). We point out that the movement of positive charge in one direction is electrically equivalent to the displacement of negative charge in the opposite direction. Therefore, the ATP8A2 current signal was associated with ATP-dependent translocation of the negatively charged PS toward the outside of the proteoliposomes (the ATP8A2 cytoplasmic side facing the external aqueous solution) [32].

It is worth noting that PC, also present in the proteoliposomes, is not a substrate for ATP8A2 and has an electrically neutral head group. As mentioned above, PC does not stimulate ATP8A2 dephosphorylation; however, the enzyme can be phosphorylated by ATP in the absence of PS and PE and in the presence of PC alone [108]. Interestingly, an ATP concentration jump on ATP8A2 reconstituted in proteoliposomes containing 100% PC yielded no significant transient current, indicating that enzyme phosphorylation by ATP ($E_1 \rightarrow E_1P$ reaction step) does not generate any electrical signal [32].

Useful information was also provided by ATP concentration jump experiments on proteoliposomes containing selected mutants of ATP8A2 [32]. In particular, the E198Q mutation was examined. Glu-198 is located in the DGET motif of the cytoplasmic A-domain of ATP8A2 that facilitates dephosphorylation of the phosphorylated intermediate. It was reported that phosphorylation by ATP is allowed in E198Q, while dephosphorylation is blocked with resulting E_2P accumulation [108]. The absence of an electrical current upon an ATP concentration jump on proteoliposomes containing E198Q indicated that the electrogenicity of ATP8A2 is not related to the $E_1 \rightarrow E_1P$ step (phosphorylation by ATP) or with the $E_1P \rightarrow E_2P$ conformational transition of the enzyme, which is in agreement with the experiments on 100% PC proteoliposomes. This finding is important to address the issue of whether ion countertransport occurs from the cytoplasmic to the exoplasmic side of the phospholipid flippase. In fact, it was shown that no charged substrate is countertransported during the $E_1 \rightarrow E_1P \rightarrow E_2P$ reaction sequence [32], thereby distinguishing the P4-ATPase from the P2-type ATPases, which transport ions in opposite directions and are therefore referred to as antiporters.

It was concluded that the electrogenicity of ATP8A2 is related to a step in the ATPase transport cycle directly involved in translocation of the phospholipid [32]: dephosphorylation of the E_2P intermediate, activated by lipid binding from the exoplasmic side, and/or the subsequent $E_2 \rightarrow E_1$ conformational transition of the dephosphoenzyme, which is associated with release of the lipid to the cytoplasmic side [104]. It is noteworthy that the findings of the SSM study of the mammalian phospholipid flippase [32] support the view that the P4-ATPase is most likely an electrogenic uniporter, as also recently reported for a bacterial Cu^+-transporting P1B-ATPase [82].

4. Conclusions

The charge transport mechanism of various P-type ATPases has been conveniently investigated by adsorbing the membrane-bound ATPase on the SSM and by activating the enzyme with a concentration jump of a suitable substrate. The transient current observed with the SSM method is a direct measurement of charge movements occurring during reactions involved in the transport cycle of the ATPase. This kind of measurements allows the identification of electrogenic partial reactions, which in turn can be used to localize charge translocation in the reaction cycle of the ion/lipid pump. It was shown that charge displacement in P-type ATPases is associated with transitions between different conformational states that facilitate the binding or release of the charged substrate.

The SSM method can provide useful information about the transport activity of P-type ATPases, as well as the molecular mechanisms of ATPase regulation, as discussed in the case of the SERCA enzyme. This is actually a very interesting research topic, which has yet to be examined in detail for some P-type ATPases, such as mammalian P4-ATPases [89]. However, also for the well-characterized P2-type ATPases Na^+,K^+-ATPase, and SERCA, a complexity of regulatory mechanisms has emerged [4,19], which requires further detailed investigation. We think that the SSM method can be usefully employed to characterize the interaction of P-type ATPases with specific regulatory partners, which can be small molecules, soluble proteins, transmembrane peptides, or the surrounding lipid bilayer [110].

Another interesting field of application of the SSM technique is related to the drug discovery process. It has been shown that drug/protein interactions can be conveniently monitored on SSMs. In particular, the SSM technique has demonstrated its usefulness to investigate the effects of various pharmaceutically relevant compounds on P-type ATPases [29,33,111], which are important targets for a variety of drugs [112–114]. Therefore, this technique represents a robust and reliable assay in drug development and evaluation studies on membrane transport proteins, as it can be used to quantify the effectiveness and potency of drugs directed toward specific protein targets, and to characterize novel drug candidates.

Funding: This research received no external funding.

Acknowledgments: Fondazione CR Firenze ID 2018.0886 for financial support.

Conflicts of Interest: The author declares no conflict of interest.

References

1. Kühlbrandt, W. Biology, structure and mechanism of P-type ATPases. *Nat. Rev. Mol. Cell Biol.* **2004**, *5*, 282–295. [CrossRef] [PubMed]

2. Bublitz, M.; Morth, J.P.; Nissen, P. P-type ATPases at a glance. *J. Cell Sci.* **2011**, *124*, 2515–2519. [CrossRef] [PubMed]

3. Palmgren, M.G.; Nissen, P. P-type ATPases. *Annu. Rev. Biophys.* **2011**, *40*, 243–266. [CrossRef] [PubMed]

4. Dyla, M.; Kjærgaard, M.; Poulsen, H.; Nissen, P. Structure and mechanism of P-type ATPase ion pumps. *Annu. Rev. Biochem.* **2020**, *89*, 583–603. [CrossRef] [PubMed]

5. Diaz, D.; Clarke, R.J. Evolutionary analysis of the lysine-rich N-terminal cytoplasmic domains of the gastric H^+,K^+-ATPase and the Na^+,K^+-ATPase. *J. Membr. Biol.* **2018**, *251*, 653–666. [CrossRef]

6. Albers, R.W. Biochemical aspects of active transport. *Ann. Rev. Biochem.* **1967**, *36*, 727–756. [CrossRef]

7. Post, R.L.; Hegyvary, C.; Kume, S. Activation by adenosine triphosphate in the phosphorylation kinetics of sodium and potassium ion transport adenosine triphosphatase. *J. Biol. Chem.* **1972**, *247*, 6530–6540.

8. Apell, H.J. Structure-function relationship in P-type ATPases—A biophysical approach. *Rev. Physiol. Biochem. Pharmacol.* **2003**, *150*, 1–35. [CrossRef]

9. Bublitz, M.; Poulsen, H.; Morth, J.P.; Nissen, P. In and out of the cation pumps: P-type ATPase structure revisited. *Curr. Opin. Struct. Biol.* **2010**, *20*, 431–439. [CrossRef]

10. Toyoshima, C.; Cornelius, F. New crystal structures of PII-type ATPases: Excitement continues. *Curr. Opin. Struct. Biol.* **2013**, *23*, 507–514. [CrossRef]

11. De Meis, L.; Vianna, A.L. Energy interconversion by the Ca^{2+}-dependent ATPase of the sarcoplasmic reticulum. *Annu. Rev. Biochem.* **1979**, *48*, 275–292. [CrossRef]

12. Toyoshima, C.; Nakasako, M.; Nomura, H.; Ogawa, H. Crystal structure of the calcium pump of sarcoplasmic reticulum at 2.6 Å resolution. *Nature* **2000**, *405*, 647–655. [CrossRef] [PubMed]

13. Toyoshima, C.; Inesi, G. Structural basis of ion pumping by Ca^{2+}-ATPase of the sarcoplasmic reticulum. *Annu. Rev. Biochem.* **2004**, *73*, 269–292. [CrossRef] [PubMed]

14. Toyoshima, C. Structural aspects of ion pumping by Ca^{2+}-ATPase of sarcoplasmic reticulum. *Arch. Biochem. Biophys.* **2008**, *476*, 3–11. [CrossRef]

15. Toyoshima, C. How Ca^{2+}-ATPase pumps ions across the sarcoplasmic reticulum membrane. *Biochim. Biophys. Acta* **2009**, *1793*, 941–946. [CrossRef] [PubMed]

16. Møller, J.V.; Olesen, C.; Winther, A.-M.L.; Nissen, P. The sarcoplasmic Ca^{2+}-ATPase: Design of a perfect chemi-osmotic pump. *Q. Rev. Biophys.* **2010**, *43*, 501–566. [CrossRef]

17. Bublitz, M.; Musgaard, M.; Poulsen, H.; Thøgersen, L.; Olesen, C.; Schiøtt, B.; Morth, J.P.; Møller, J.V.; Nissen, P. Ion pathways in the sarcoplasmic reticulum Ca^{2+}-ATPase. *J. Biol. Chem.* **2013**, *288*, 10759–10765. [CrossRef] [PubMed]

18. Dyla, M.; Basse Hansen, S.; Nissen, P.; Kjaergaard, M. Structural dynamics of P-type ATPase ion pumps. *Biochem. Soc. Trans.* **2019**, *47*, 1247–1257. [CrossRef]

19. Aguayo-Ortiz, R.; Espinoza-Fonseca, L.M. Linking biochemical and structural states of SERCA: Achievements, challenges and new opportunities. *Int. J. Mol. Sci.* **2020**, *21*, 4146. [CrossRef]

20. Tadini-Buoninsegni, F.; Bartolommei, G.; Moncelli, M.R.; Fendler, K. Charge transfer in P-type ATPases investigated on planar membranes. *Arch. Biochem. Biophys.* **2008**, *476*, 75–86. [CrossRef]

21. Seifert, K.; Fendler, K.; Bamberg, E. Charge transport by ion translocating membrane proteins on solid supported membranes. *Biophys. J.* **1993**, *64*, 384–391. [CrossRef]

22. Pintschovius, J.; Fendler, K. Charge Translocation by the Na^+/K^+-ATPase investigated on solid supported membranes: Rapid solution exchange with a new technique. *Biophys. J.* **1999**, *76*, 814–826. [CrossRef]

23. Bartolommei, G.; Moncelli, M.R.; Rispoli, G.; Kelety, B.; Tadini-Buoninsegni, F. Electrogenic ion pumps investigated on a solid supported membrane: Comparison of current and voltage measurements. *Langmuir* **2009**, *25*, 10925–10931. [CrossRef]

24. Schulz, P.; Garcia-Celma, J.J.; Fendler, K. SSM-based electrophysiology. *Methods* **2008**, *46*, 97–103. [CrossRef]

25. Tadini-Buoninsegni, F.; Fendler, K. Recording of pump and transporter activity using solid-supported membranes (SSM-based electrophysiology). In *Pumps, Channels and Transporters: Methods of Functional Analysis*; Clarke, R.J., Khalid, M.A.A., Eds.; John Wiley & Sons Inc.: Hoboken, NJ, USA, 2015; pp. 147–177. [CrossRef]

26. Pintschovius, J.; Fendler, K.; Bamberg, E. Charge translocation by the Na^+/K^+-ATPase investigated on solid supported membranes: Cytoplasmic cation binding and release. *Biophys. J.* **1999**, *76*, 827–836. [CrossRef]

27. Tadini-Buoninsegni, F.; Bartolommei, G.; Moncelli, M.R.; Guidelli, R.; Inesi, G. Pre-steady state electrogenic events of Ca^{2+}/H^+ exchange and transport by the Ca^{2+}-ATPase. *J. Biol. Chem.* **2006**, *281*, 37720–37727. [CrossRef]

28. Liu, Y.; Pilankatta, R.; Lewis, D.; Inesi, G.; Tadini-Buoninsegni, F.; Bartolommei, G.; Moncelli, M.R. High-yield heterologous expression of wild type and mutant Ca^{2+} ATPase: Characterization of Ca^{2+} binding sites by charge transfer. *J. Mol. Biol.* **2009**, *391*, 858–871. [CrossRef] [PubMed]

29. Kelety, B.; Diekert, K.; Tobien, J.; Watzke, N.; Dörner, W.; Obrdlik, P.; Fendler, K. Transporter assays using solid supported membranes: A novel screening platform for drug discovery. *Assay Drug Dev. Technol.* **2006**, *4*, 575–582. [CrossRef]

30. Mattle, D.; Zhang, L.; Sitsel, O.; Pedersen, L.T.; Moncelli, M.R.; Tadini-Buoninsegni, F.; Gourdon, P.; Rees, D.C.; Nissen, P.; Meloni, G. A sulfur-based transport pathway in Cu^+-ATPases. *EMBO Rep.* **2015**, *16*, 728–740. [CrossRef]

31. Tadini-Buoninsegni, F.; Bartolommei, G.; Moncelli, M.R.; Pilankatta, R.; Lewis, D.; Inesi, G. ATP dependent charge movement in ATP7B Cu^+-ATPase is demonstrated by pre-steady state electrical measurements. *FEBS Lett.* **2010**, *584*, 4619–4622. [CrossRef]

32. Tadini-Buoninsegni, F.; Mikkelsen, S.A.; Mogensen, L.S.; Molday, R.S.; Andersen, J.P. Phosphatidylserine flipping by the P4-ATPase ATP8A2 is electrogenic. *Proc. Natl. Acad. Sci. USA* **2019**, *116*, 16332–16337. [CrossRef] [PubMed]

33. Tadini-Buoninsegni, F.; Palchetti, I. Label-Free Bioelectrochemical methods for evaluation of anticancer drug effects at a molecular level. *Sensors* **2020**, *20*, 1812. [CrossRef] [PubMed]

34. Geibel, S.; Flores-Herr, N.; Licher, T.; Vollert, H. Establishment of cell-free electrophysiology for ion transporters: Application for pharmacological profiling. *J. Biomol. Screen.* **2006**, *11*, 262–268. [CrossRef]

35. Barthmes, M.; Liao, J.; Jiang, Y.; Brüggemann, A.; Wahl-Schott, C. Electrophysiological characterization of the archaeal transporter NCX_Mj using solid supported membrane technology. *J. Gen. Physiol.* **2016**, *147*, 485–496. [CrossRef] [PubMed]

36. Bazzone, A.; Barthmes, M.; Fendler, K. SSM-based electrophysiology for transporter research. *Methods Enzymol.* **2017**, *594*, 31–83. [CrossRef] [PubMed]

37. Inesi, G.; Tadini-Buoninsegni, F. Ca^{2+}/H^+ exchange, lumenal Ca^{2+} release and Ca^{2+}/ATP coupling ratios in the sarcoplasmic reticulum ATPase. *J. Cell Commun. Signal.* **2014**, *8*, 5–11. [CrossRef]

38. Primeau, J.O.; Armanious, G.P.; Fisher, M.E.; Young, H.S. The SarcoEndoplasmic Reticulum Calcium ATPase. *Subcell. Biochem.* **2018**, *87*, 229–258. [CrossRef]

39. Chen, J.; Sitsel, A.; Benoy, V.; Sepúlveda, M.R.; Vangheluwe, P. Primary active Ca^{2+} transport systems in health and disease. *Cold Spring Harb. Perspect. Biol.* **2020**, *12*, a035113. [CrossRef]

40. Brini, M.; Carafoli, E. Calcium pumps in health and disease. *Physiol. Rev.* **2009**, *89*, 1341–1378. [CrossRef]

41. Grover, A.K.; Khan, I. Calcium pump isoforms: Diversity, selectivity and plasticity. *Cell Calcium* **1992**, *13*, 9–17. [CrossRef]

42. Periasamy, M.; Kalyanasundaram, A. SERCA pump isoforms: Their role in calcium transport and disease. *Muscle Nerve* **2007**, *35*, 430–442. [CrossRef] [PubMed]

43. Franzini-Armstrong, C.; Ferguson, D.G. Density and disposition of Ca^{2+}-ATPase in sarcoplasmic reticulum membrane as determined by shadowing techniques. *Biophys. J.* **1985**, *48*, 607–615. [CrossRef]

44. Tadini-Buoninsegni, F.; Bartolommei, G.; Moncelli, M.R.; Tal, D.M.; Lewis, D.; Inesi, G. Effects of high-affinity inhibitors on partial reactions, charge movements, and conformational states of the Ca^{2+} transport ATPase (sarco-endoplasmic reticulum Ca^{2+} ATPase). *Mol. Pharmacol.* **2008**, *73*, 1134–1140. [CrossRef] [PubMed]

45. Sagara, Y.; Inesi, G. Inhibition of the sarcoplasmic reticulum Ca^{2+} transport ATPase by thapsigargin at subnanomolar concentrations. *J. Biol. Chem.* **1991**, *266*, 13503–13506.

46. Lytton, J.; Westlin, M.; Hanley, M.R. Thapsigargin inhibits the sarcoplasmic or endoplasmic reticulum Ca-ATPase family of calcium pumps. *J. Biol. Chem.* **1991**, *266*, 17067–17071.

47. Inesi, G.; Lewis, D.; Toyoshima, C.; Hirata, A.; de Meis, L. Conformational fluctuations of the Ca^{2+}-ATPase in the native membrane environment. Effects of pH, temperature, catalytic substrates, and thapsigargin. *J. Biol. Chem.* **2008**, *283*, 1189–1196. [CrossRef]

48. Lewis, D.; Pilankatta, R.; Inesi, G.; Bartolommei, G.; Moncelli, M.R.; Tadini-Buoninsegni, F. Distinctive features of catalytic and transport mechanisms in mammalian sarco-endoplasmic reticulum Ca^{2+} ATPase (SERCA) and Cu^+ (ATP7A/B) ATPases. *J. Biol. Chem.* **2012**, *287*, 32717–32727. [CrossRef]

49. Yu, X.; Hao, L.; Inesi, G. A pK change of acidic residues contributes to cation countertransport in the Ca-ATPase of sarcoplasmic reticulum. Role of H^+ in Ca^{2+}-ATPase countertransport. *J. Biol. Chem.* **1994**, *269*, 16656–16661.

50. Cornelius, F.; Møller, J.V. Electrogenic pump current of sarcoplasmic reticulum Ca^{2+}-ATPase reconstituted at high lipid/protein ratio. *FEBS Lett.* **1991**, *284*, 46–50. [CrossRef]

51. Levy, D.; Seigneuret, M.; Bluzat, A.; Rigaud, J.L. Evidence for proton countertransport by the sarcoplasmic reticulum Ca^{2+}-ATPase during calcium transport in reconstituted proteoliposomes with low ionic permeability. *J. Biol. Chem.* **1990**, *265*, 19524–19534.

52. Yu, X.; Carroll, S.; Rigaud, J.L.; Inesi, G. H^+ countertransport and electrogenicity of the sarcoplasmic reticulum Ca^{2+} pump in reconstituted proteoliposomes. *Biophys. J.* **1993**, *64*, 1232–1242. [CrossRef]

53. MacLennan, D.H.; Kranias, E.G. Phospholamban: A crucial regulator of cardiac contractility. *Nat. Rev. Mol. Cell Biol.* **2003**, *4*, 566–577. [CrossRef] [PubMed]

54. Shaikh, S.A.; Sahoo, S.K.; Periasamy, M. Phospholamban and sarcolipin: Are they functionally redundant or distinct regulators of the Sarco (Endo)Plasmic Reticulum Calcium ATPase? *J. Mol. Cell. Cardiol.* **2016**, *91*, 81–91. [CrossRef] [PubMed]

55. Glaves, J.P.; Trieber, C.A.; Ceholski, D.K.; Stokes, D.L.; Young, H.S. Phosphorylation and mutation of phospholamban alter physical interactions with the sarcoplasmic reticulum calcium pump. *J. Mol. Biol.* **2011**, *405*, 707–723. [CrossRef]

56. Glaves, J.P.; Primeau, J.O.; Espinoza-Fonseca, L.M.; Lemieux, M.J.; Young, H.S. The phospholamban pentamer alters function of the sarcoplasmic reticulum calcium pump SERCA. *Biophys. J.* **2019**, *116*, 633–647. [CrossRef]

57. Smeazzetto, S.; Armanious, G.P.; Moncelli, M.R.; Bak, J.J.; Lemieux, M.J.; Young, H.S.; Tadini-Buoninsegni, F. Conformational memory in the association of the transmembrane protein phospholamban with the sarcoplasmic reticulum calcium pump SERCA. *J. Biol. Chem.* **2017**, *292*, 21330–21339. [CrossRef]

58. Schörner, M.; Beyer, S.R.; Southall, J.; Cogdell, R.J.; Köhler, J. Conformational memory of a protein revealed by single-molecule spectroscopy. *J. Phys. Chem. B* **2015**, *119*, 13964–13970. [CrossRef]

59. Csermely, P.; Kunsic, N.; Mendik, P.; Kerestély, M.; Faragó, T.; Veres, D.V.; Tompa, P. Learning of signaling networks: Molecular mechanisms. *Trends Biochem. Sci.* **2020**, *45*, 284–294. [CrossRef]

60. Anderson, D.M.; Anderson, K.M.; Chang, C.L.; Makarewich, C.A.; Nelson, B.R.; McAnally, J.R.; Kasaragod, P.; Shelton, J.M.; Liou, J.; Bassel-Duby, R.; et al. A micropeptide encoded by a putative long noncoding RNA regulates muscle performance. *Cell* **2015**, *160*, 595–606. [CrossRef]

61. Anderson, D.M.; Makarewich, C.A.; Anderson, K.M.; Shelton, J.M.; Bezprozvannaya, S.; Bassel-Duby, R.; Olson, E.N. Widespread control of calcium signaling by a family of SERCA-inhibiting micropeptides. *Sci. Signal.* **2016**, *9*, ra119. [CrossRef]

62. Nelson, B.R.; Makarewich, C.A.; Anderson, D.M.; Winders, B.R.; Troupes, C.D.; Wu, F.; Reese, A.L.; McAnally, J.R.; Chen, X.; Kavalali, E.T.; et al. A peptide encoded by a transcript annotated as long noncoding RNA enhances SERCA activity in muscle. *Science* **2016**, *351*, 271–275. [CrossRef] [PubMed]

63. Singh, D.R.; Dalton, M.P.; Cho, E.E.; Pribadi, M.P.; Zak, T.J.; Šeflová, J.; Makarewich, C.A.; Olson, E.N.; Robia, S.L. Newly discovered micropeptide regulators of SERCA form oligomers but bind to the pump as monomers. *J. Mol. Biol.* **2019**, *431*, 4429–4443. [CrossRef] [PubMed]

64. Lutsenko, S.; Barnes, N.L.; Bartee, M.Y.; Dmitriev, O.Y. Function and regulation of human copper-transporting ATPases. *Physiol. Rev.* **2007**, *87*, 1011–1046. [CrossRef] [PubMed]

65. Kaplan, J.H.; Lutsenko, S. Copper transport in mammalian cells: Special care for a metal with special needs. *J. Biol. Chem.* **2009**, *284*, 25461–25465. [CrossRef]

66. La Fontaine, S.; Ackland, M.L.; Mercer, J.F. Mammalian copper-transporting P-type ATPases, ATP7A and ATP7B: Emerging roles. *Int. J. Biochem. Cell Biol.* **2010**, *42*, 206–209. [CrossRef]

67. Gourdon, P.; Sitsel, O.; Lykkegaard Karlsen, J.; Birk Møller, L.; Nissen, P. Structural models of the human copper P-type ATPases ATP7A and ATP7B. *Biol. Chem.* **2012**, *393*, 205–216. [CrossRef]

68. Petruzzelli, R.; Polishchuk, R.S. Activity and trafficking of copper-transporting ATPases in tumor development and defense against platinum-based drugs. *Cells* **2019**, *8*, 1080. [CrossRef]

69. Yu, C.H.; Dolgova, N.V.; Dmitriev, O.Y. Dynamics of the metal binding domains and regulation of the human copper transporters ATP7B and ATP7A. *IUBMB Life* **2017**, *69*, 226–235. [CrossRef]

70. Argüello, J.M.; González-Guerrero, M.; Raimunda, D. Bacterial transition metal P1B-ATPases: Transport mechanism and roles in virulence. *Biochemistry* **2011**, *50*, 9940–9949. [CrossRef]

71. Argüello, J.M.; Raimunda, D.; González-Guerrero, M. Metal transport across biomembranes: Emerging models for a distinct chemistry. *J. Biol. Chem.* **2012**, *287*, 13510–13517. [CrossRef]

72. Mattle, D.; Sitsel, O.; Autzen, H.E.; Meloni, G.; Gourdon, P.; Nissen, P. On allosteric modulation of P-type Cu$^+$-ATPases. *J. Mol. Biol.* **2013**, *425*, 2299–2308. [CrossRef]

73. Sitsel, O.; Grønberg, C.; Autzen, H.E.; Wang, K.; Meloni, G.; Nissen, P.; Gourdon, P. Structure and function of Cu(I)- and Zn(II)-ATPases. *Biochemistry* **2015**, *54*, 5673–5683. [CrossRef] [PubMed]

74. Wittung-Stafshede, P. Unresolved questions in human copper pump mechanisms. *Q. Rev. Biophys.* **2015**, *48*, 471–478. [CrossRef] [PubMed]

75. Gourdon, P.; Liu, X.Y.; Skjørringe, T.; Morth, J.P.; Møller, L.B.; Pedersen, B.P.; Nissen, P. Crystal structure of a copper-transporting PIB-type ATPase. *Nature* **2011**, *475*, 59–64. [CrossRef] [PubMed]

76. Andersson, M.; Mattle, D.; Sitsel, O.; Klymchuk, T.; Nielsen, A.M.; Møller, L.B.; White, S.H.; Nissen, P.; Gourdon, P. Copper-transporting P-type ATPases use a unique ion-release pathway. *Nat. Struct. Mol. Biol.* **2014**, *21*, 43–48. [CrossRef] [PubMed]

77. Tadini-Buoninsegni, F.; Bartolommei, G.; Moncelli, M.R.; Inesi, G.; Galliani, A.; Sinisi, M.; Losacco, M.; Natile, G.; Arnesano, F. Translocation of platinum anticancer drugs by human copper ATPases ATP7A and ATP7B. *Angew. Chem. Int. Ed. Engl.* **2014**, *53*, 1297–1301. [CrossRef]

78. Verjovski-Almeida, S.; Kurzmack, M.; Inesi, G. Partial reactions in the catalytic and transport cycle of sarcoplasmic reticulum ATPase. *Biochemistry* **1978**, *17*, 5006–5013. [CrossRef]

79. González-Guerrero, M.; Eren, E.; Rawat, S.; Stemmler, T.L.; Argüello, J.M. Structure of the two transmembrane Cu$^+$ transport sites of the Cu$^+$-ATPases. *J. Biol. Chem.* **2008**, *283*, 29753–29759. [CrossRef]

80. Obara, K.; Miyashita, N.; Xu, C.; Toyoshima, I.; Sugita, Y.; Inesi, G.; Toyoshima, C. Structural role of countertransport revealed in Ca^{2+} pump crystal structure in the absence of Ca^{2+}. *Proc. Natl. Acad. Sci. USA* **2005**, *102*, 14489–14496. [CrossRef]

81. Wang, K.; Sitsel, O.; Meloni, G.; Autzen, H.E.; Andersson, M.; Klymchuk, T.; Nielsen, A.M.; Rees, D.C.; Nissen, P.; Gourdon, P. Structure and mechanism of Zn^{2+}-transporting P-type ATPases. *Nature* **2014**, *514*, 518–522. [CrossRef]

82. Abeyrathna, N.; Abeyrathna, S.; Morgan, M.T.; Fahrni, C.J.; Meloni, G. Transmembrane Cu(I) P-type ATPase pumps are electrogenic uniporters. *Dalton Trans.* **2020**. [CrossRef] [PubMed]

83. Holthuis, J.C.; Levine, T.P. Lipid traffic: Floppy drives and a superhighway. *Nat. Rev. Mol. Cell Biol.* **2005**, *6*, 209–220. [CrossRef]

84. Lenoir, G.; Williamson, P.; Holthuis, J.C. On the origin of lipid asymmetry: The flip side of ion transport. *Curr. Opin. Chem. Biol.* **2007**, *11*, 654–661. [CrossRef] [PubMed]

85. Nagata, S.; Sakuragi, T.; Segawa, K. Flippase and scramblase for phosphatidylserine exposure. *Curr. Opin. Immunol.* **2020**, *62*, 31–38. [CrossRef] [PubMed]

86. Clarke, R.J.; Hossain, K.R.; Cao, K. Physiological roles of transverse lipid asymmetry of animal membranes. *Biochim. Biophys. Acta Biomembr.* **2020**, *1862*, 183382. [CrossRef] [PubMed]

87. Puts, C.F.; Holthuis, J.C. Mechanism and significance of P4 ATPase-catalyzed lipid transport: Lessons from a Na$^+$/K$^+$-pump. *Biochim. Biophys. Acta* **2009**, *1791*, 603–611. [CrossRef] [PubMed]

88. Hankins, H.M.; Baldridge, R.D.; Xu, P.; Graham, T.R. Role of flippases, scramblases and transfer proteins in phosphatidylserine subcellular distribution. *Traffic* **2015**, *16*, 35–47. [CrossRef]

89. Andersen, J.P.; Vestergaard, A.L.; Mikkelsen, S.A.; Mogensen, L.S.; Chalat, M.; Molday, R.S. P4-ATPases as phospholipid flippases-structure, function, and enigmas. *Front. Physiol.* **2016**, *7*, 275. [CrossRef]

90. Bevers, E.M.; Williamson, P.L. Getting to the Outer Leaflet: Physiology of phosphatidylserine exposure at the plasma membrane. *Physiol. Rev.* **2016**, *96*, 605–645. [CrossRef]

91. Devaux, P.F.; López-Montero, I.; Bryde, S. Proteins involved in lipid translocation in eukaryotic cells. *Chem. Phys. Lipids* **2006**, *141*, 119–132. [CrossRef]

92. Van der Mark, V.A.; Elferink, R.P.; Paulusma, C.C. P4 ATPases: Flippases in health and disease. *Int. J. Mol. Sci.* **2013**, *14*, 7897–7922. [CrossRef] [PubMed]

93. Lopez-Marques, R.L.; Theorin, L.; Palmgren, M.G.; Pomorski, T.G. P4-ATPases: Lipid flippases in cell membranes. *Pflugers Arch.* **2014**, *466*, 1227–1240. [CrossRef] [PubMed]

94. Panatala, R.; Hennrich, H.; Holthuis, J.C. Inner workings and biological impact of phospholipid flippases. *J. Cell Sci.* **2015**, *128*, 2021–2032. [CrossRef] [PubMed]

95. Roland, B.P.; Graham, T.R. Decoding P4-ATPase substrate interactions. *Crit. Rev. Biochem. Mol. Biol.* **2016**, *51*, 513–527. [CrossRef]

96. Bryde, S.; Hennrich, H.; Verhulst, P.M.; Devaux, P.F.; Lenoir, G.; Holthuis, J.C. CDC50 proteins are critical components of the human class-1 P4-ATPase transport machinery. *J. Biol. Chem.* **2010**, *285*, 40562–40572. [CrossRef]

97. Coleman, J.A.; Molday, R.S. Critical role of the beta-subunit CDC50A in the stable expression, assembly, subcellular localization, and lipid transport activity of the P4-ATPase ATP8A2. *J. Biol. Chem.* **2011**, *286*, 17205–17216. [CrossRef]

98. Timcenko, M.; Lyons, J.A.; Januliene, D.; Ulstrup, J.J.; Dieudonné, T.; Montigny, C.; Ash, M.R.; Karlsen, J.L.; Boesen, T.; Kühlbrandt, W.; et al. Structure and autoregulation of a P4-ATPase lipid flippase. *Nature* **2019**, *571*, 366–370. [CrossRef]

99. Bai, L.; Kovach, A.; You, Q.; Hsu, H.C.; Zhao, G.; Li, H. Autoinhibition and activation mechanisms of the eukaryotic lipid flippase Drs2p-Cdc50p. *Nat. Commun.* **2019**, *10*, 4142. [CrossRef]

100. Hiraizumi, M.; Yamashita, K.; Nishizawa, T.; Nureki, O. Cryo-EM structures capture the transport cycle of the P4-ATPase flippase. *Science* **2019**, *365*, 1149–1155. [CrossRef]

101. Lyons, J.A.; Timcenko, M.; Dieudonné, T.; Lenoir, G.; Nissen, P. P4-ATPases: How an old dog learnt new tricks-structure and mechanism of lipid flippases. *Curr. Opin. Struct. Biol.* **2020**, *63*, 65–73. [CrossRef]

102. Nakanishi, H.; Irie, K.; Segawa, K.; Hasegawa, K.; Fujiyoshi, Y.; Nagata, S.; Abe, K. Crystal structure of a human plasma membrane phospholipid flippase. *J. Biol. Chem.* **2020**, *295*, 10180–10194. [CrossRef]

103. Baldridge, R.D.; Graham, T.R. Two-gate mechanism for phospholipid selection and transport by type IV P-type ATPases. *Proc. Natl. Acad. Sci. USA* **2013**, *110*, E358–E367. [CrossRef] [PubMed]

104. Vestergaard, A.L.; Coleman, J.A.; Lemmin, T.; Mikkelsen, S.A.; Molday, L.L.; Vilsen, B.; Molday, R.S.; Dal Peraro, M.; Andersen, J.P. Critical roles of isoleucine-364 and adjacent residues in a hydrophobic gate control of phospholipid transport by the mammalian P4-ATPase ATP8A2. *Proc. Natl. Acad. Sci. USA* **2014**, *111*, E1334–E1343. [CrossRef] [PubMed]

105. Montigny, C.; Lyons, J.; Champeil, P.; Nissen, P.; Lenoir, G. On the molecular mechanism of flippase- and scramblase-mediated phospholipid transport. *Biochim. Biophys. Acta* **2016**, *1861*, 767–783. [CrossRef] [PubMed]

106. Jensen, M.S.; Costa, S.R.; Duelli, A.S.; Andersen, P.A.; Poulsen, L.R.; Stanchev, L.D.; Gourdon, P.; Palmgren, M.; Günther Pomorski, T.; López-Marqués, R.L. Phospholipid flipping involves a central cavity in P4 ATPases. *Sci. Rep.* **2017**, *7*, 17621. [CrossRef] [PubMed]

107. Gantzel, R.H.; Mogensen, L.S.; Mikkelsen, S.A.; Vilsen, B.; Molday, R.S.; Vestergaard, A.L.; Andersen, J.P. Disease mutations reveal residues critical to the interaction of P4-ATPases with lipid substrates. *Sci. Rep.* **2017**, *7*, 10418. [CrossRef]

108. Coleman, J.A.; Vestergaard, A.L.; Molday, R.S.; Vilsen, B.; Andersen, J.P. Critical role of a transmembrane lysine in aminophospholipid transport by mammalian photoreceptor P4-ATPase ATP8A2. *Proc. Natl. Acad. Sci. USA* **2012**, *109*, 1449–1454. [CrossRef]

109. Coleman, J.A.; Kwok, M.C.; Molday, R.S. Localization, purification, and functional reconstitution of the P4-ATPase Atp8a2, a phosphatidylserine flippase in photoreceptor disc membranes. *J. Biol. Chem.* **2009**, *284*, 32670–32679. [CrossRef]

110. Hossain, K.R.; Clarke, R.J. General and specific interactions of the phospholipid bilayer with P-type ATPases. *Biophys. Rev.* **2019**, *11*, 353–364. [CrossRef]

111. Tadini-Buoninsegni, F.; Smeazzetto, S.; Gualdani, R.; Moncelli, M.R. Drug interactions with the Ca^{2+}-ATPase from sarco(endo)plasmic reticulum (SERCA). *Front. Mol. Biosci.* **2018**, *5*, 36. [CrossRef]

112. Yatime, L.; Buch-Pedersen, M.J.; Musgaard, M.; Morth, J.P.; Winther, A.-M.L.; Pedersen, B.P.; Olesen, C.; Andersen, J.P.; Vilsen, B.; Schiøtt, B.; et al. P-type ATPases as drug targets: Tools for medicine and science. *Biochim. Biophys. Acta-Bioenerg.* **2009**, *1787*, 207–220. [CrossRef]

113. Michelangeli, F.; East, J.M. A diversity of SERCA Ca^{2+} pump inhibitors. *Biochem. Soc. Trans.* **2011**, *39*, 789–797. [CrossRef] [PubMed]

114. Peterková, L.; Kmoníčková, E.; Ruml, T.; Rimpelová, S. Sarco/endoplasmic reticulum calcium ATPase inhibitors: Beyond anticancer perspective. *J. Med. Chem.* **2020**, *63*, 1937–1963. [CrossRef] [PubMed]

MDPI

St. Alban-Anlage 66

4052 Basel

Switzerland

Tel. +41 61 683 77 34

Fax +41 61 302 89 18

www.mdpi.com

Molecules Editorial Office

E-mail: molecules@mdpi.com

www.mdpi.com/journal/molecules